Alexander Carmele

Theory for strongly coupled quantum dot cavity quantum electrodynamics

Alexander Carmele

Theory for strongly coupled quantum dot cavity quantum electrodynamics

Photon statistics and phonon signatures in quantum light emission

Südwestdeutscher Verlag für Hochschulschriften

Impressum/Imprint (nur für Deutschland/only for Germany)
Bibliografische Information der Deutschen Nationalbibliothek: Die Deutsche Nationalbibliothek verzeichnet diese Publikation in der Deutschen Nationalbibliografie; detaillierte bibliografische Daten sind im Internet über http://dnb.d-nb.de abrufbar.
Alle in diesem Buch genannten Marken und Produktnamen unterliegen warenzeichen-, marken- oder patentrechtlichem Schutz bzw. sind Warenzeichen oder eingetragene Warenzeichen der jeweiligen Inhaber. Die Wiedergabe von Marken, Produktnamen, Gebrauchsnamen, Handelsnamen, Warenbezeichnungen u.s.w. in diesem Werk berechtigt auch ohne besondere Kennzeichnung nicht zu der Annahme, dass solche Namen im Sinne der Warenzeichen- und Markenschutzgesetzgebung als frei zu betrachten wären und daher von jedermann benutzt werden dürften.

Verlag: Südwestdeutscher Verlag für Hochschulschriften GmbH & Co. KG
Dudweiler Landstr. 99, 66123 Saarbrücken, Deutschland
Telefon +49 681 37 20 271-1, Telefax +49 681 37 20 271-0
Email: info@svh-verlag.de

Approved by: Berlin, TU, Diss., 2010

Herstellung in Deutschland:
Schaltungsdienst Lange o.H.G., Berlin
Books on Demand GmbH, Norderstedt
Reha GmbH, Saarbrücken
Amazon Distribution GmbH, Leipzig
ISBN: 978-3-8381-2847-4

Imprint (only for USA, GB)
Bibliographic information published by the Deutsche Nationalbibliothek: The Deutsche Nationalbibliothek lists this publication in the Deutsche Nationalbibliografie; detailed bibliographic data are available in the Internet at http://dnb.d-nb.de.
Any brand names and product names mentioned in this book are subject to trademark, brand or patent protection and are trademarks or registered trademarks of their respective holders. The use of brand names, product names, common names, trade names, product descriptions etc. even without a particular marking in this works is in no way to be construed to mean that such names may be regarded as unrestricted in respect of trademark and brand protection legislation and could thus be used by anyone.

Publisher: Südwestdeutscher Verlag für Hochschulschriften GmbH & Co. KG
Dudweiler Landstr. 99, 66123 Saarbrücken, Germany
Phone +49 681 37 20 271-1, Fax +49 681 37 20 271-0
Email: info@svh-verlag.de

Printed in the U.S.A.
Printed in the U.K. by (see last page)
ISBN: 978-3-8381-2847-4

Copyright © 2011 by the author and Südwestdeutscher Verlag für Hochschulschriften GmbH & Co. KG and licensors
All rights reserved. Saarbrücken 2011

Contents

1 Introduction   **3**
1.1 Motivation 3
1.2 Highlights 5
1.3 Structure 5

2 Quantum optics in the equation of motion approach: The factorization problem   **7**
2.1 Jaynes-Cummings Hamiltonian 8
2.2 Jaynes-Cummings solution 8
2.3 Hierarchy problem and cluster expansion 10

3 Quantum dot cavity quantum electrodynamics   **16**
3.1 Mathematical induction approach 16
    3.1.1 The Hamiltonian 16
    3.1.2 General equations of motion 17
    3.1.3 Initial conditions 21
3.2 Benchmarking the induction model 22
    3.2.1 Independent boson model:LO-phonon satellite peaks 22
    3.2.2 Jaynes-Cummings model (1): Vacuum Rabi splitting 24
    3.2.3 Jaynes-Cummings model (2): Collapse and revivals 27
    3.2.4 Jaynes-Cummings model (3): Analytical solution 29
3.3 LO-phonon QD cavity-QED: LO-phonon assisted cavity feeding . . . 31
    3.3.1 Initial conditions 31
    3.3.2 Equations of Motion 32
    3.3.3 LO-phonon QD cavity-QED: Optical absorption spectrum . . 32
    3.3.4 Phonon-induced Rabi frequency modification 35
3.4 LO-phonon QD cavity-QED: Degradation of photon-statistics 37
    3.4.1 Initial conditions and equations of motion 37
    3.4.2 LO-phonon induced anti-bunching 38
3.5 LO-phonon QD cavity-QED: Biexciton cascade 44
    3.5.1 Electron-electron interaction 44
    3.5.2 Electron-photon interaction 46
    3.5.3 Biexciton cascade: Equations of motion 48
    3.5.4 Biexciton cascade: Dynamics in the strong coupling regime 50

3.5.5	Biexciton cascade and entanglement	53
3.5.6	Biexciton cascade: Weak coupling regime	55

4 Quantum dot - wetting layer cavity quantum electrodynamics — 61
- 4.1 Standard cluster expansion (SCE) beyond the one-electron assumption (OEA) 63
  - 4.1.1 Hartree-Fock approximation within the SCE 63
  - 4.1.2 Modified equations of motion within the SCE 65
- 4.2 Photon probability cluster expansion approach (PPCE) 67
  - 4.2.1 Photon probabilities expansion 68
  - 4.2.2 PPCE and photon-statistics 69
  - 4.2.3 PPCE and Hartree Fock factorization 71
  - 4.2.4 PPCE and modified equations of motion 74
- 4.3 Photon dynamics in the PPCE and SCE approach 75
  - 4.3.1 Intensity-intensity correlations within the Hartree-Fock approximation 75
  - 4.3.2 Pauli-blocking dependent Rabi oscillation amplitude 78
- 4.4 PPCE and environment coupling 80
  - 4.4.1 Electrical pumping 81
  - 4.4.2 β-factor 82
  - 4.4.3 Cavity loss 85
  - 4.4.4 Quantum dot - wetting layer laser equations 87
- 4.5 Electrically pumped single photon emitter 87
  - 4.5.1 Laser dynamics of an electrically driven QD 88
  - 4.5.2 Parameter studies of a single QD laser device 92
  - 4.5.3 Atom and QD-WL rate equations in the single-photon limit 95

5 Conclusion and outlook — 98

6 Appendix — 101
- 6.1 Numerical parameters 101
- 6.2 Biexciton cascade in the strong coupling regime: Equations of motions 101
- 6.3 Biexciton cascade in the weak coupling regime: Equations of motions 104
- 6.4 Derivation of the Hartree-Fock factorization 106
- 6.5 Photon probability picture 108
- 6.6 Jaynes-Cummings solution in the photon probability picture 109
- 6.7 Modified photon-assisted polarization (electron picture) 111
- 6.8 Cavity loss in the photon probability picture 111

Danksagung — 116

Bibliography — 117

1 Introduction

1.1 Motivation

Quantum optical properties pave the way to new and spectacular technological advances by improving the accuracy of measurements done so far with stabilized lasers [Shi07], increasing the degree of security in quantum cryptography via the knowledge of quantum states [UFT07] or exploiting entanglement for quantum computing [BPM+97]. In addition to an atom-photon interface, ideal sources for deterministic quantum light emission with tunable photon statis-

Figure 1.1: Micropillar SEM [SDT+05].

tics are semiconductor quantum dots (QDs) coupled to an optical microcavity, cf. Fig. 1.1 [STH+09; FMR+09]. These systems also exhibit signatures of strong coupling [RSL+04]. In this regime, the electron-photon interaction outrivals the dissipation processes and distinct quantum optical phenomena become accessible, such as vacuum Rabi oscillations or the vacuum Rabi splitting [KGK+06; MKH08]. In this limit, the quantized nature of light is of primary importance, and semiconductor research begins to shed new light on well-known quantum optical effects, e.g., gigantic, experimentally robust resonances of vacuum Rabi splitting of the second rung in the Jaynes-Cummings ladder [KK08; SKK08]. Conceptually new approaches are introduced, such as implementing the coherent control of an exciton two-level system (qubit) by means of a time-dependent electric interaction [VSSM+10]. For weak coupling, a smooth transition from strong photon bunching to Poissonian statistics in the output characteristics of semiconductor microcavity lasers is observed [GWJ08].

One of the most feasible features of solid state environments is the capability of positioning nanostructures such as quantum dots permanently in a microcavity [KKKG06]. The size and geometry of the QD control the properties, e.g., the coupling strength and the confinement energies [Sti01]. Those systems are designated candidates for future technological applications, including single-photon emitters [LO05; MIM+00]. In optical quantum-information processing, semiconductor quantum dots inside a microcavity have emerged as promising sources of polarization-entangled photon pairs [TPT06]. However, the limits of future technological applications depend on understanding of quantum many-body phenomena in condensed matter environments such as electron-electron scattering or phonon-induced dephasing [SCK01; MZ07; MR09; RCB+09; DMK05], which are not present in atom-photon systems.

Common non-Markovian quantum mechanical approaches for treating the electron-photon interaction result in a hierarchy problem, if the Jaynes-Cummings-Model (JCM), which is valid only in the case of uncoupled two-level-systems interacting with a single cavity mode [JC63], is exceeded. Hence, quantum optics in semiconductors, where Coulomb-interaction and electron-phonon interaction take place, calls for different approaches such as the Born approximation [VWW01] or the cluster-expansion [KK08]. Recent work incorporates several aspects of many-body phenomena into the calculation of semiconductor quantum dot cavity quantum electrodynamics (QD-CQED): Absorption spectra in the presence of strong electron-phonon interaction have been studied with polaron operator techniques [WRI02; MKH08]. The excitonic time dynamics

Figure 1.2: Time-resolved emission scheme, calculated in the equation of motion approach [DMR+10].

in a perturbation approach, based on the Lamb-Dicke approximation [LMC08] for LA-phonons, comparable to the second-order Born approximation, has been theoretically investigated. Often, many-particle interactions are taken into account via an equation of motion approach [GMK06; MTK07]. However, factorization techniques become problematic in case of strongly correlated interactions, such as the electron-photon interaction in the strong coupling and single-photon regime.

This thesis focuses on the theoretical description of semiconductor QD-CQED beyond the JCM, including non-Markovian contributions and many-particle interactions such as electron-phonon coupling on a microscopical level. The equation of motion approach is chosen as an established technique to explain recent experimental results and to give insight into the underlying physics. Two theoretical frameworks are introduced to provide a simulation tool for parameter studies in the single-photon limit: a mathematical induction approach and the photon-probability cluster expansion. In this regime, quantum correlations are important, and fully quantized approaches become necessary, which take into account the complete and strongly correlated quantum dynamics of the observables of interest: the photon density, the intensity-intensity correlation $g^{(2)}(t, 0)$-function, the polarization and the electron density. The mathematical induction model solves the QD-CQED dynamics including the electron-longitudinal optical (LO) phonon interaction up to an arbitrary accuracy in case of a fixed number of electrons in the system but including dissipation processes such as a cavity loss or a radiative decay. In the presence of a wetting layer, the number of carriers, electrons and holes, is not fixed and a modified Hartree-Fock factorization based on the photon-probability cluster expansion (PPCE) approach is introduced to treat the occurring many-particle correlations.

1.2 Highlights

The mathematical induction approach allows the inclusion of the LO-phonon interaction up to an arbitrary order. As a result, the LO-phonon assisted QD-CQED dynamics is computed and a novel feature predicted: *Antibunching of Thermal Radiation by a Room-Temperature Phonon Bath* [CRCK10]. Furthermore, LO-phonon assisted cavity feeding leads to interesting anharmonic Rabi oscillations for elevated temperatures. In the weak and strong coupling regime, the biexciton cascade in a QD is computed and for the first time *a temperature dependent analysis of the formation dynamics of an entangled photon pair* is derived [CMD+10]. For wetting layer contributions in the QD-CQED, a *few-photon model of the optical emission of semiconductor quantum dots* is introduced, the PPCE [RCSK09a]. A modified Hartree-Fock factorization is derived and the *photon statistics of a single quantum dot in a microcavity* is calculated [SRK+10; SCR+10], and hereby the completion of a *theory of few photon dynamics in electrically pumped light emitting QD devices* provided [CRD+10; DMR+10].

1.3 Structure

The thesis is divided into five chapters. The introduction in this chapter 1 contains the motivation, the highlights and the outline of the thesis. The fundamental quantum optical model to study cavity-QED, the Jaynes-Cummings model (JCM), is discussed in chapter 2. The JCM provides an analytical solution under the condition of an isolated two-level system with one electron interacting with a single cavity mode without dissipation processes. These conditions are not fulfilled in the case of semiconductor QD-CQED. In order to consider further interactions, such as electron-phonon coupling, and losses due to environmental coupling, the equation of motion approach is chosen, which allows the extension of the JCM. However, this leads to the well-known hierarchy problem and factorization approaches become necessary. In chapter 3, an equation of motion approach is introduced, which does not rely on factorization and solves the hierarchy problem via a mathematical induction model. The induction approach goes beyond the JCM by including the LO-phonon interaction up to arbitrary accuracy and cavity losses. However, similiar to the JCM, the induction model relies on the assumption of a fixed number of electrons in the system. In chapter 4, a Hartree-Fock based perturbation approach is introduced and is valid without assuming a fixed number of carriers. In particular, this is important in studies of a single-QD laser in the presence of wetting layer (WL) carriers, since the WL leads to a modified spontaneous emission source term. In chapter 5, a conclusion and outlook is given.

1 Introduction

2 Quantum optics in the equation of motion approach: The factorization problem

Theoretical approaches in semiconductor many-particle physics are often based on an equation of motion scheme in the formalism of second quantization [HK04]. The equations of motion of observables are derived via commutating them with the full Hamiltonian of the investigated system [KJHK99]. In contrast to typical master equation approaches, the equation of motion approach does not rely on the Markovian approximation, i.e., the neglection of memory effects [Car99]. Therefore, to study quantum correlations of higher order or ultrafast phenomena, the equation of motion approach is advantageous and enables the investigation of novel quantum features, such as excitation induced dephasing [SCK01; SCK04], Coulomb scattering processes in quantum well and microcavity systems [KKM01; KK08] or many-particle-induced non-Lorentzian lineshapes in semiconductor nano-optics [FAD+02; MKH08].

The non-Markovian equation of motion approach results in the hierarchy problem. As an example, the quantum correlations of second order couple to quantum correlations of the third order [KJHK99]. The set of equations of motion is not closed, unless the quantum correlations are factorized at a given level. A typical approach is the cluster expansion or correlation expansion method [Fri96; GWLJ07; KK08]. Via a complete factorization of occurring quantum correlations, the set of equations of motion is truncated and casted into a closed set, which is numerically solvable. However, this cluster expansion approach is problematic for systems, in which the quantum correlations are in the order of magnitude of the corresponding mean value of the observable [RCSK09a]. A prominent example is the single-photon emitter, where the fluctuations and mean value are of the same order of magnitude.

In the following chapter, this factorization problem is discussed. A model Hamiltonian of a two-level system strongly coupled to a single cavity mode is assumed. This simplest possible fully quantized model of interest is exactly solvable [ENSM80]. The solution is called the Jaynes-Cummings model (JCM) [JC63] and is taken as a benchmark for the solutions derived in the equation of motion approach. The deviations from the exact Jaynes-Cummings model are discussed.

2.1 Jaynes-Cummings Hamiltonian

The Hamiltonian of a single two-level system interacting with a single-mode field reads in rotating-wave [HK04] and dipole approximation [SZ97]:

$$H = \hbar\omega_v a_v^\dagger a_v + \hbar\omega_c a_c^\dagger a_c + \hbar\omega_0 c^\dagger c - \hbar M(a_v^\dagger a_c c^\dagger + a_c^\dagger a_v c), \quad (2.1)$$

where ω_0 is the frequency of the quantized light field with bosonic creation ($c^\dagger$) and annihilation operators (c). In the two-level system one electron is assumed, either in the ground state (v) with energy $\hbar\omega_v$ or in the excited state (c) with $\hbar\omega_c$ the excited-state energy. The fermionic creation (annhiliation) operator denote $a_{c/v}^{(\dagger)}$ and M is the electron-photon coupling element.

Here, the coupling element is assumed to be real and includes the polarization of the photon and the dipole strength of the two-level system, as well as the corresponding frequency of the single-mode light field. The interaction is depicted in Fig. 2.1. The two-level system is driven by the quantized light field. If the electron is excited via a photon absorption from the ground state to the excited state ($a_c^\dagger a_v$), a photon from the quantized light field is annihilated (c). If the electron relaxes from the excited state to the ground state via spontaneous or induced emission ($a_v^\dagger a_c$), a photon is created ($c^\dagger$).

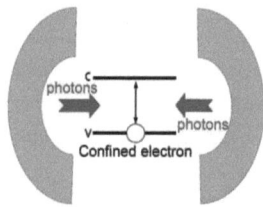

Figure 2.1: JCM Hamiltonian

As long as the interaction is restricted to the electron-photon case, other processes are not possible. Losses, such as cavity loss or radiative decay or pure dephasing, are not considered. The zero point energy in the free-energy part of the photon-field $\sum_k \hbar\omega_k/2$ is omitted for convenience [SZ97]. The zero-point energy sums frequencies without an upper bound and is accordingly infinite, an awkward feature of the quantized light field. Fortunately, usual measurements in quantum optics do not apply apparatus, which would register a response proportional to the zero-point energy. They response only to the change in the total energy of the electromagnetic field [Lou83].

2.2 Jaynes-Cummings solution

In the case of a two-level system with one electron interacting with a quantized light field, the dynamics can be solved exactly. Several approaches are possible, e.g., the unitary time-evolution operator method [SZ97]. If the excited and ground state energy are chosen as: $\hbar\omega_c = \frac{\hbar\tilde{\omega}}{2} = -\hbar\omega_v$, Eq. (2.1) equals the Jaynes-Cummings Hamiltonian and reads:

$$H = \underbrace{\frac{\hbar\tilde{\omega}}{2}(a_c^\dagger a_c - a_v^\dagger a_v)}_{H_I} + \underbrace{\hbar\omega_0 c^\dagger c - \hbar M \left(a_v^\dagger a_c c^\dagger + a_c^\dagger a_v c \right)}_{H_{II}}. \quad (2.2)$$

The dynamics of the system depends only on H_{II}. It is convenient to describe the dynamics in the interaction picture. The transformed interaction Hamiltonian H_{II} reads:

$$H_{II}^{int} = e^{\frac{i}{\hbar}H_I t} H_{II} e^{-\frac{i}{\hbar}H_I t}. \tag{2.3}$$

Using the one-electron assumption, i.e., any combination of four electron operators becomes zero in the expectation value and can be neglected ($\langle a_i^\dagger a_j^\dagger a_k a_l \rangle = 0$ for $i,j,k,l = v,c$), and assuming the two-level system to be in resonance with the light field, H_{II} reads explicitely:

$$H_{II}^{int} = -\hbar M \left(a_v^\dagger a_c c^\dagger + a_c^\dagger a_v c \right). \tag{2.4}$$

The time evolution of the system for a given state $|\Psi(t)\rangle$ is calculated via the unitary time evolution operator and reads $|\Psi(t)\rangle = U(t)|\Psi(0)\rangle$ with $|\Psi(0)\rangle$ as the initial state and the unitary time-evolution operator in the interaction picture:

$$U(t) = e^{-\frac{i}{\hbar}H_{II}^{int} t}. \tag{2.5}$$

After evaluating the exponential function, again using the one-electron assumption, the time-evolution operator reads:

$$U(t) = \cos\left(M\sqrt{c^\dagger c + 1}\, t\right) a_c^\dagger a_c + \cos\left(M\sqrt{c^\dagger c}\, t\right) a_v^\dagger a_v$$
$$-i \frac{\sin\left(M\sqrt{c^\dagger c + 1}\, t\right)}{\sqrt{c^\dagger c + 1}} c\, a_c^\dagger a_v - i c^\dagger \frac{\sin\left(M\sqrt{c^\dagger c + 1}\, t\right)}{\sqrt{c^\dagger c + 1}} a_v^\dagger a_c. \tag{2.6}$$

Now, the time evolution of a given state is easily calculated, bearing in mind, that cosine and the square root are analytical functions and can be expressed in a Taylor series. As an example, the electron is initially in the excited state with N photons in the system, i.e. the state is $|\Psi(0)\rangle = |c,N\rangle$. After applying the unitary time evolution operator, the time evolution reads:

$$|\Psi(t)\rangle = U(t)|c,N\rangle = \cos\left(M\sqrt{N+1}\,t\right)|c,N\rangle - i\sin\left(M\sqrt{N+1}\,t\right)|v,N+1\rangle. \tag{2.7}$$

The analytical solution results from a division of the Hilbert space in subspaces, consisting of the only two occurring processes in the system: 1) a photon is absorbed ($N \to N-1$) and the electron moves into the excited state $|v\rangle \to |c\rangle$ and 2) a photon is emitted via spontaneous or induced emission ($N \to N+1$) and the electron moves back into the ground state $|c\rangle \to |v\rangle$. The time evolution of ground and excited state densities is calculated via the probability amplitudes given in the time evolution of the system state with $C_c^N = \langle c,N|\Psi(t)\rangle$ and $C_v^N = \langle v,N+1|\Psi(t)\rangle$. The exact solution reads:

$$\langle a_v^\dagger a_v \rangle = |C_v^N(t)|^2 = \sin^2(\Omega_N t), \qquad \langle a_c^\dagger a_c \rangle = |C_c^N(t)|^2 = \cos^2(\Omega_N t), \tag{2.8}$$

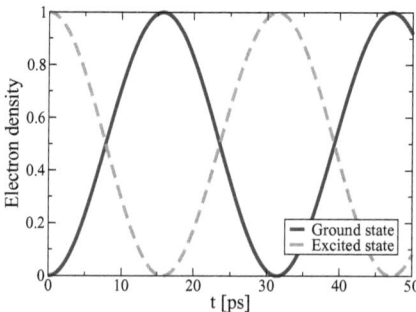

Figure 2.2: Analytical solution of the JCM for a two-level system interacting with a quantized light field. Rabi oscillations occur between the ground and excited state.

with the quantum-mechanical Rabi frequency $\Omega_N = M\sqrt{N+1}$, cf. Fig. 2.2. The solution includes quantum optical properties of emission processes, e.g. spontaneous emission, seen in the vacuum Rabi oscillations ($N = 0$) and experimentally observed in [NRS+94]. The solution enables one to implement photon statistics dependent Rabi oscillations by calculating a superposition of JCM solutions [LP07; SZ97]:

$$\langle a_c^\dagger a_c \rangle = \sum_{n=0}^{\infty} |p_n|^2 \cos^2(\Omega_n t),$$

where $|p_n|^2$ is the probability to find n photons in a given mode within a given photon mean value. Although the JCM is a powerful tool, to implement quantum optical properties in the emission and absorption dynamics, it is more or less limited to an isolated two-level system. Including, for example, Coulomb or phonon induced dephasing processes microscopically, the interaction Hamiltonian is more complex and the presented analytical solution Eq. (2.8) is not valid.

2.3 Hierarchy problem and cluster expansion

In this section, a simple example of the hierarchy problem is presented, which occurs in systems with many-particle quantum correlations. This equation of motion approach is chosen to include non-Markovian many-particle features in the calculations. Starting from the Hamiltonian Eq. (2.1), equations of motion for the observables of interest are derived via the Heisenberg equation of motion:

$$-i\hbar \partial_t \, a_i^\dagger a_j = [\hat{H}, a_i^\dagger a_j]. \tag{2.9}$$

2 Quantum Optics in the Equation of Motion Approach: The Factorization Problem

A single-mode quantized light field and one electron in a two-level system are considered, as well as resonance between the light mode ω_0 and band gap frequency $\omega_{cv} = \omega_c - \omega_v$ assumed. The dynamics of the photon density, of the excited-, and ground state read:

$$\partial_t \langle a_c^\dagger a_c \rangle = 2\,\mathrm{Im}\left[M \langle a_v^\dagger a_c c^\dagger \rangle \right], \qquad (2.10)$$

$$\partial_t \langle a_v^\dagger a_v \rangle = -2\,\mathrm{Im}\left[M \langle a_v^\dagger a_c c^\dagger \rangle \right] = \partial_t \langle c^\dagger c \rangle \qquad (2.11)$$

The equations of motion for the photon density and the ground state density are identical. This is easy to understand: the JCM considers a closed system with particle conservation, i.e., the number of excitations and photons (ν) is constant: $\langle \nu \rangle = \langle a_c^\dagger a_c \rangle + \langle c^\dagger c \rangle$ = const. Assuming an initially excited two-level system without photons in the cavity, $\langle \nu \rangle = 1$ is valid for all times. The one-electron assumption is also valid for all times $\langle a_c^\dagger a_c \rangle + \langle a_v^\dagger a_v \rangle = 1$, which yields for the conservation of particles: $\langle \nu \rangle = 1 = 1 - \langle a_v^\dagger a_v \rangle(t) + \langle c^\dagger c \rangle(t) \rightarrow \langle a_v^\dagger a_v \rangle(t) = \langle c^\dagger c \rangle(t)$. Of course, this is only valid for zero photons in the cavity and an initially excited two-level system. The equations of motion of the densities couple to the photon-assisted polarization:

$$\partial_t \langle a_v^\dagger a_c c^\dagger \rangle = -iM \langle a_c^\dagger a_c \rangle - iM \left(\langle a_c^\dagger a_c c^\dagger c \rangle - \langle a_v^\dagger a_v c^\dagger c \rangle \right). \qquad (2.12)$$

The hierarchy problem is now obvious: a quantity with n photon operators couples to a quantity with $n + 1$ photon operators [GWLJ07]. If the one-electron assumption is abandoned, another hierarchy problem occurs within the electronic system, cf. Chap. 4. The set of differential equations must be truncated to obtain a closed system with respect to the electron-photon coupling M. Common approaches rely on the Born approximation [RBSK07; VWW01] or the cluster expansion [FMWS97; KJHK99]. Here, the cluster expansion is applied to derive the truncated form for Eq. (2.12):

$$\partial_t \langle a_v^\dagger a_c c^\dagger \rangle = -iM \langle a_c^\dagger a_c \rangle - iM \langle c^\dagger c \rangle \left(\langle a_c^\dagger a_c \rangle - \langle a_v^\dagger a_v \rangle \right)$$
$$-iM \left(\langle a_c^\dagger a_c c^\dagger c \rangle^c - \langle a_v^\dagger a_v c^\dagger c \rangle^c \right). \qquad (2.13)$$

Pure, coherently driven quantities such as $\langle a_v^\dagger a_c \rangle, \langle c^\dagger \rangle$ are neglected. They do not contribute to the combined electron-photon dynamics without the inclusion of a classical pump field [RCSK09a]. If the correction terms $\langle a_i^\dagger a_i c^\dagger c \rangle^c$ are neglected, the set of differential equations [Eq. (2.10) -(2.11) and (2.13)] is closed and can be solved numerically and analytically. With the same initial condition: $\langle c^\dagger c \rangle(0) = 0$ and $\langle a_c^\dagger a_c \rangle(0) = 1$, the solution is plotted in Fig. 2.3. Again, Rabi-oscillations are visible for the JCM (solid, green line). The second-order correlation expansion is a good approximation for the first pico seconds. After 10 ps, the deviation becomes larger and the approximation fails completely. Negative probabilities occur due to the factorization. Dissipation processes (cavity loss or electronic dephasing) can prevent negativities but do not solve the underlying problem. In conclusion, a second-order correlation expansion in the electron-light coupling element is not sufficient to factorize the combined electron-photon dynamics correctly. Higher order correlation terms

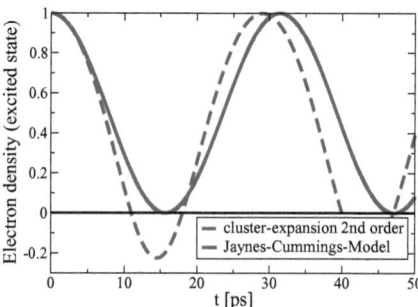

Figure 2.3: Comparison of the second-order cluster-expansion (dashed, red line) in the electron-photon coupling element with the exact solution of the JCM (solid, green line). After 10 ps, the deviation of the exact solution becomes very large.

in Eq. (2.13) are necessary. The dynamics of the correction terms are derived via the same approach. The photon-density assisted electron density in the excited state reads:

$$\partial_t \langle a_c^\dagger a_c c^\dagger c \rangle = 2 \operatorname{Im}\left[M \langle a_v^\dagger a_c c^\dagger c^\dagger c \rangle \right]. \quad (2.14)$$

The correlation term is derived by:

$$\partial_t \langle a_c^\dagger a_c c^\dagger c \rangle^c = \partial_t \langle a_c^\dagger a_c c^\dagger c \rangle - \partial_t \left[\langle a_c^\dagger a_c \rangle \langle c^\dagger c \rangle + \langle a_c^\dagger a_c c^\dagger \rangle \langle c \rangle + \langle a_c^\dagger a_c c \rangle \langle c^\dagger \rangle \right].$$

After the neglecting coherently driven contributions, the equation of motion reads

$$\partial_t \langle a_c^\dagger a_c c^\dagger c \rangle^c = 2 \operatorname{Im}[M \langle a_v^\dagger a_c c^\dagger c^\dagger c \rangle^c + M \langle a_v^\dagger a_c c^\dagger \rangle^c (\langle c^\dagger c \rangle + \langle a_c^\dagger a_c \rangle)]. \quad (2.15)$$

Using the one-electron assumption and Eq. (2.15), $\partial_t \langle a_c^\dagger a_c c^\dagger c \rangle^c = -\partial_t \langle a_v^\dagger a_v c^\dagger c \rangle^c$ is shown. This identity is disturbing, since ground state and excited state dynamics differs in quantum optical schemes. The ground state is driven by the higher-order photon-assisted polarization and additionally by the lower-order photon-assisted polarization, whereas the excited state couples only to the higher-order photon-assisted polarization. In the cluster expansion approach, photon density-assisted ground and excited state lose this significant difference in third order perturbation theory and are described via the same equation of motion, only with different sign and initial conditions. Neglecting again the higher-order correlation term in Eq. (2.15), the set of equations of motion is closed and can be evaluated numerically. The solution is plotted in Fig. 2.4(a). A truncation at the third order is problematic within a equation of motion approach. In Fig. 2.4(a), no negativities occur, but the oscillation amplitude of the cluster expansion solution (dashed, red line) is too small. Furthermore, the Rabi oscillation frequency is wrong due to the decreased amplitude. Note, that the deviation from the JCM solution is bigger (difference of the minima: 0.3), compared to the solution of the second order correlation expansion (−0.2). An alternating convergence to the JCM is improbable. This becomes even clearer,

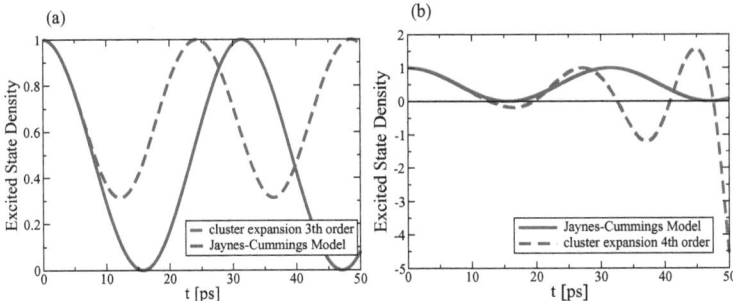

Figure 2.4: (a) Comparison of the third order cluster expansion and the JCM solution. Although the approximation is quite good in the first 10 ps, the deviation becomes very large at later times. (b) Solution of the 4th order cluster expansion (red, dashed line) and the JCM solution (green, solid line). The solution obtained via the cluster expansion diverges.

when investigating the 4th order solution in the cluster expansion.

The higher-order photon-assisted polarization includes the intensity-intensity correlation implicitly, and after the factorization, explicitly. The dynamics in the fourth order of the electron-photon coupling element reads:

$$\partial_t \langle a_v^\dagger a_c c^\dagger c^\dagger c \rangle = -2i M \langle a_c^\dagger a_c c^\dagger c \rangle - iM \left(\langle a_c^\dagger a_c c^\dagger c^\dagger c \, c \rangle - \langle a_v^\dagger a_v c^\dagger c^\dagger c \, c \rangle \right), \quad (2.16)$$

$$\partial_t \langle c^\dagger c^\dagger cc \rangle = -4 \operatorname{Im} \left[\bar{M} \langle a_v^\dagger a_c c^\dagger c^\dagger c \rangle \right]. \quad (2.17)$$

The factorization implies more correction terms than in the lower order due to the different possibilities to factorize singlett, dublett, and triplett expectation values:

$$\partial_t \langle a_v^\dagger a_c c^\dagger c^\dagger c \rangle^c = \quad (2.18)$$
$$= -2iM \langle a_c^\dagger a_c c^\dagger c \rangle^c - iM \langle c^\dagger c^\dagger c \, c \rangle^c \left(\langle a_c^\dagger a_c \rangle - \langle a_v^\dagger a_v \rangle \right)$$
$$- 2iM \langle c^\dagger c \rangle \left(\langle a_c^\dagger a_c c^\dagger c \rangle^c - \langle a_v^\dagger a_v c^\dagger c \rangle^c \right) + 4 \langle a_v^\dagger a_c c^\dagger \rangle^c \operatorname{Im} \left[M \langle a_v^\dagger a_c c^\dagger \rangle^c \right].$$

This equation yields unexpected terms, proportional to the photon density and results even in a diverging dynamics. In Fig. 2.4(b), the solution of the 4th order cluster expansion (red, dashed line) and the JCM solution (green, solid line) are plotted. Up to 10 ps, the factorized solution reproduces the JCM. For longer times, the factorized solution deviates strongly from the JCM solution. In the 4th order the solution additionally loses the periodicity property. The excited state density finally diverges.

The cluster expansion factorization approach leads to wrong results in the single-photon limit: few emitters are correlated with few photons in the cavity.

For an ensemble of emitters, at least 10 – 50, or for a high number of photons (25 – 50), the cluster expansion yields good results [SKK08; GWJ08; KK08]. In particular, the second order factorization converges to the solution of the JCM model. In Fig. 2.5, the Jaynes-Cummings model solution

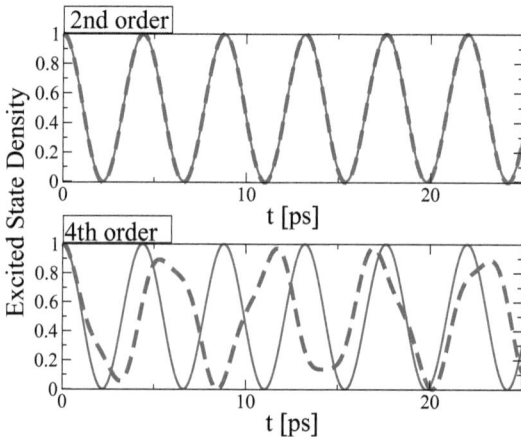

Figure 2.5: Solution of the 2nd and 4th order cluster expansion (red, dashed line) and the JCM solution (green, solid line) for a high number of photons ($N = 50$).

(green, solid line) is compared with the *2nd*-order (upper panel) and *4th*-order (lower panel) solution, obtained via factorization in the cluster expansion scheme for the initial condition of one excited two-level system with initially $N = 50$ photons in the cavity. The *2nd*-order solution fits the JCM solution perfectly. The deviations are vanishingly small, which proves the cluster expansion scheme to be well suited for a situation, where one emitter interacts with a large number of photons. In other words, the electron and photon dynamics are weakly correlated, since only one out of 50 photons is absorbed or re-emitted. Assuming this weak correlation regime, the *2nd*-order equations of motion can be rewritten. Given the one electron approximation, Eq. (2.12) is written as:

$$\partial_t \langle a_v^\dagger a_c c^\dagger \rangle = -iM \langle a_c^\dagger a_c \rangle - iM \langle c^\dagger c \rangle (2 \langle a_c^\dagger a_c \rangle - 1).$$

For a large photon number N, $\langle c^\dagger c \rangle = N$ dominates over the product $\langle a_c^\dagger a_c \rangle \langle c^\dagger c \rangle$. In the case of a single emitter, the photon density fluctuates ±1 around N, whereas the product of excited state and photon number oscillates between 0 and $2N$. The larger N, the closer the photon-assisted polarization appears in the JCM limit. However, in the single-photon limit, N and $\langle a_c^\dagger a_c \rangle N$ are of the same order of magnitude and the deviations from the JCM become large.

In Fig. 2.5 (lower panel), the solution of the *4th*-order is compared with the JCM. Again, the solution is closer to the JCM solution and shows a periodic behavior, which is lost in the few photon limit, but the deviations are larger in comparison to the deviation between the JCM solution and the *2nd*-order solution. This suggests, that higher order correlation expansion terms do not correspond and alter the dynamics artificially, introducing anharmonic oscillation behavior seen for, e.g., $t = 6$ ps. The factorization scheme must be adapted to the situation of weakly coupled electrons and photons.

E.g., the cluster expansion is done only within the photon field and correlation terms containing both fermion and boson contributions such as $\langle a_c^\dagger a_c c^\dagger c \rangle^c \approx 0$ are neglected:

$$\langle a_c^\dagger a_c c^\dagger c^\dagger cc \rangle = \langle a_c^\dagger a_c \rangle \langle c^\dagger c^\dagger cc \rangle + 2 \langle a_c^\dagger a_c \rangle \langle c^\dagger c \rangle \langle c^\dagger c \rangle + \langle a_c^\dagger a_c c^\dagger c^\dagger cc \rangle^c. \quad (2.19)$$

Hence, correlations between the interacting fields are assumed to be small. Now, the correction term of 4th order reads:

$$\partial_t \langle a_v^\dagger a_c c^\dagger c \rangle^c = \quad (2.20)$$
$$= -2iM \langle a_c^\dagger a_c c^\dagger c \rangle^c - iM \langle c^\dagger c^\dagger c c \rangle^c \left(\langle a_c^\dagger a_c \rangle - \langle a_v^\dagger a_v \rangle \right)$$
$$+ 2iM \langle c^\dagger c \rangle \left(\langle a_c^\dagger a_c c^\dagger c \rangle^c - \langle a_v^\dagger a_v c^\dagger c \rangle^c \right) + 4 \langle a_v^\dagger a_c c^\dagger \rangle^c \, \text{Im} \left[M \langle a_v^\dagger a_c c^\dagger \rangle^c \right].$$

The Rabi oscillations, which are described exact within the JCM, are approximated to a satisfying accuracy by Eq. (2.20). Fig. 2.6 shows no deviation between the approximated solution, derived via this modified cluster expansion, and the exact solution from the JCM. However, the suggested truncation

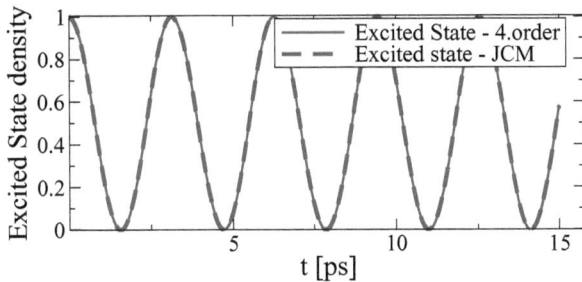

Figure 2.6: The approximated Rabi oscillation via the *modified* cluster expansion in the 4th-order of the light-coupling element are in good accordance with the exact solution given by the JCM.

scheme is not adaptable to the strong coupling regime, for few photons and few emitters. In case of weak coupling, presumely without cavity enhanced dynamics, the factorization schemes such as the cluster expansion or the Born approximation work.

3 Quantum dot cavity quantum electrodynamics

3.1 Mathematical induction approach

In this section, the QD cavity-QED within the equation of motion approach is described by a mathematical induction approach. Based on the factorization problems of the cluster expansion in the single-photon or strongly correlation limit, cf. Chap. 2, a theoretical framework is derived to treat strongly coupled quantum correlations up to an arbitrary accuracy in a non-Markovian approach. This novel approach allows one to include more interactions without any factorization, e.g. the electron-phonon interaction, or a classical pump field, or Coulomb contribution in the case of two or more electrons in the QD. If a set of general equations of motion for the involved quantum correlations is found, the induction method is limited solely by numerical means, as long as the number of electrons in the QD is fixed. In small

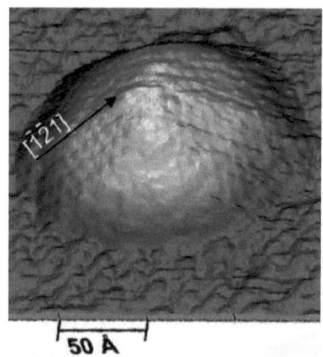

Figure 3.1: GaAs/InAs-QD [MGJ01].

and deep QDs, this is a good approximation, since the WL electrons are energetically well separated from the QD states and the QD is considered to be atom-like [SGB99a; Sti01]. For shallow QDs, the WL carriers easily scatter into the QD states and a fixed number of electrons cannot be assumed, cf. Chap. 4.

This section is organized as follows. First, the Hamiltonian is discussed in Sec. 3.1.1, including QD specific properties. In Sec. 3.1.2, a general set of equations of motion is derived and discussed, including all occurring quantum correlations. To benchmark this induction approach, its results are compared with the JCM and the independent Boson solution in Sec. 3.2. Then, the LO-phonon impact on the QD cavity-QED is evaluated with respect to the absorption spectrum and LO-phonon assisted cavity feeding in Sec. 3.3 and the remarkable feature, that a classical cavity field is transformed into a non-classical via the LO-phonon interaction, i.e. LO-phonon induced anti-bunching in Sec. 3.4.

3.1.1 The Hamiltonian

Focusing on a QD strongly coupled to a microcavity within a semiconductor bulk material, the Heisenberg equation of motion approach is chosen to calculate the full quantum kinetics of quantum optical observables of interest. The QD is populated with one electron and treated as a two level system with one valence band state with energy $\hbar\omega_v$ and one conduction band state with energy $\hbar\omega_c$, e.g., InAs/GaAs-QD with a base length of smaller than 13 nm [SGB99a] or free-standing QD fabricated in the heterogeneous droplet epitaxy method [SMO+02]. For larger QD, where the one electron assumption is not valid anymore, other approaches become necessary [RCSK09b]. The QD band gap energy $\hbar\omega_{cv} := \hbar\omega_c - \hbar\omega_v$ and the wave functions are calculated in the effective mass approximation [HK04]. The QD interacts with the classical pump field $\Omega(t)$, the effective single cavity mode ω_0 and with LO-phonons of the semiconductor bulk material, cf. Fig 3.2. The LO-phonon frequency is ω_{LO} and a constant dispersion is assumed (Einstein model) [MZ07]. The total Hamiltonian of the system reads in rotating-wave and dipole approximation [KK08; HK04]:

$$\begin{aligned} H = & \hbar\omega_v a_v^\dagger a_v + \hbar\omega_c a_c^\dagger a_c - \hbar\Omega(t)(a_v^\dagger a_c + a_c^\dagger a_v) \\ & + \hbar\omega_0 c^\dagger c - \hbar M(a_v^\dagger a_c c^\dagger + a_c^\dagger a_v c) \\ & + \hbar \sum_q \omega_{LO} b_q^\dagger b_q + a_c^\dagger a_c \left(g_q^c b_q + g_q^{c*} b_q^\dagger\right) + a_v^\dagger a_v \left(g_q^v b_q + g_q^{v*} b_q^\dagger\right), \end{aligned} \qquad (3.1)$$

where $a^{(\dagger)}$ denotes the fermionic annihilation (creation) operators for the electron in the quantum dot and the bosonic annihilation (creation) operators $c^{(\dagger)}$ for photons and $b_q^{(\dagger)}$ for LO-phonons in the mode q. Electron and photons interact via the electron-light coupling-element M, which depends on the geometry, size and position of the quantum dot in the semiconductor microcavity [RSL+04; TPT06; Hoh10]. The electron-phonon interaction is given with the exciton-phonon Fröhlich-coupling element $g_q^{(c,v)}$ [MZ07; KAK02]. The modification of the bulk phonon modes due to the geometry and size of the microcavity is assumed to be small and is therefore neglected [WLW09]. Due to the large band gap frequency ω_{cv} in comparison to the actual phonon energies in GaAs, non-diagonal electron-phonon contribution are also not included. A semi-classical electron-light interaction is included, assuming an excitation of the carriers in the QD by an external laser field. The corresponding interaction strength is determined by $\Omega(t) = ME(t)/\hbar$ with $E(t) = \tilde{E}(t) \cos(\omega_l t)$, oscillating with ω_l and including a time dependent amplitude $\tilde{E}(t)$.

In this investigation, all parameters are chosen for an InAs/GaAs-quantum dot in a GaAs semiconductor bulk material, cf. App. 6.1 for the list of numerical parameters.

3.1.2 General equations of motion

A general set of equations is derived with the Heisenberg equation of motion $-i\hbar\partial_t A = [H, A]$, where A is an arbitrary operator. For example, the excited state $\langle a_c^\dagger a_c \rangle$ dynamics reads:

$$\partial_t \langle a_c^\dagger a_c \rangle = 2 \operatorname{Im}\left(M \langle a_v^\dagger a_c c^\dagger \rangle + \Omega(t) \langle a_v^\dagger a_c \rangle\right). \qquad (3.2)$$

3 QUANTUM DOT CAVITY QUANTUM ELECTRODYNAMICS

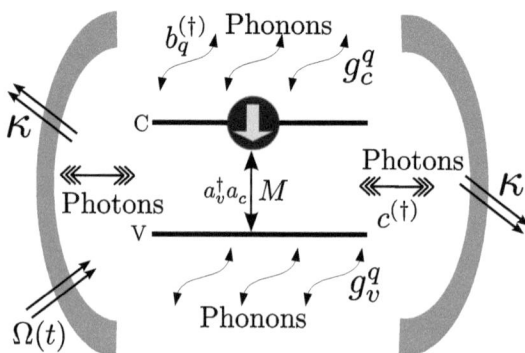

Figure 3.2: Scheme of the semiconductor quantum dot cavity-QED. The quantum dot is treated as a two-level system with one valence v and one conduction band state c. The electron $a_{v/c}^{(\dagger)}$ in the quantum dot interacts with LO-phonons $b_q^{(\dagger)}$ from the semiconductor bulk material via the electron-phonon coupling element $g_q^{(v,c)}$ and with the photons $c^{(\dagger)}$ of the cavity via the electron-photon coupling element M. A phenomenologically included cavity loss κ is considered, as well as a classical pump field $\Omega(t)$.

This equation couples to the photon-assisted polarization via the electron-photon coupling element M and to the polarization via the classical pump field $\Omega(t)$. The photon-assisted polarization is given by:

$$\begin{aligned}\partial_t \langle a_v^\dagger a_c c^\dagger \rangle &= -i(\omega_{cv} - \omega - i\kappa)\langle a_v^\dagger a_c c^\dagger \rangle - i\Omega(t)\left(\langle a_c^\dagger a_c c^\dagger \rangle - \langle a_v^\dagger a_v c^\dagger \rangle\right) \\ &\quad -i M \langle a_v^\dagger a_c \rangle - i M \left(\langle a_c^\dagger a_c c^\dagger c \rangle - \langle a_v^\dagger a_v c^\dagger c \rangle\right) \\ &\quad -i \sum_q (g_q^c - g_q^v) \langle a_v^\dagger a_c c^\dagger b \rangle - i \sum_q (g_q^{c*} - g_q^{v*}) \langle a_v^\dagger a_c c^\dagger b^\dagger \rangle. \end{aligned}$$ (3.3)

To model a realistic microcavity, a cavity-loss is introduced phenomenologically, which is assumed to be $\kappa = 10\,\mu eV$ [RSL+04]. The photon-assisted polarization couples again to higher order photon-assisted quantities such as $\langle a_c^\dagger a_c c^\dagger c \rangle$. Additionally, the LO-phonon operators enter into the dynamics. The hierarchy problem is obvious. Here, a numerically exactly solvabel model is introduced in order to calculate the LO-phonon quantum cavity-QED for an initially fixed number of electrons in the QD [CRCK10].

In the first step, a general set of equations is derived via mathematical induction. It is not necessary, to calculate all interactions at once. Applying the product rule for phonon- and photon assisted quantities, the interaction adds up:

$$\partial_t \left(a_c^\dagger a_c c^\dagger c b_q^\dagger b_q\right) = \left(\partial_t\, a_c^\dagger a_c c^\dagger c\right) b_q^\dagger b_q + c^\dagger c \left(\partial_t\, a_c^\dagger a_c b_q^\dagger b_q\right).$$ (3.4)

Therefore, the general equations of motion are derived separately for the phonon- and photon-assisted quantities and added afterwards. Here, it is particularly convenient, since the electron-photon coupling is non-diagonal, whereas the electron-phonon coupling is diagonal in the Hamiltonian. The

3 Quantum Dot Cavity Quantum Electrodynamics

derivation of a general set of equations of motion is straightforward, and proven via mathematical induction. To illustrate the derivation, an example is given: the generalized equation of motion for photon-correlations. The equation of motion of the photon-density, or first-order photon correlation, and intensity-intensity correlation, or second-order photon correlation read:

$$\partial_t(c^\dagger c) = iM\left(a_v^\dagger a_c c^\dagger - a_c^\dagger a_v c\right),$$
$$\partial_t(c^\dagger c^\dagger cc) = 2\,iM\left(a_v^\dagger a_c c^\dagger c^\dagger c - a_c^\dagger a_v c^\dagger cc\right).$$

In this particularly simple example, the general equation of motion is easy to guess with p an integer:

$$\partial_t(c^{\dagger p} c^p) = p\,iM\left(a_v^\dagger a_c c^{\dagger p} c^{p-1} - a_c^\dagger a_v c^{\dagger p-1} c^p\right). \tag{3.5}$$

It follows from the commutator algebra $[A, F(B)] = [A, B]F'(B)$ a convenient rule [CT99; AHK99], which is applied to arbitrary powers of the photon-operators:

$$\partial_t c^{\dagger p} = \partial_t F(c^\dagger) = -iM[a_v^\dagger a_c c^\dagger + a_c^\dagger a_v c, F(c^\dagger)] = -iMa_c^\dagger a_v[c, c^\dagger]F'(c^\dagger) = -ipMa_c^\dagger a_v c^{\dagger p-1},$$

and Eq. (3.5) is proven and is valid for the hermetian conjugate as well. In this illustrated derivation, the combined electron-phonon and electron-photon dynamics are computed and expressed in a general set of equations of motion. For convenience, some abbreviations are introduced, taken into account the constant dispersion of the LO-phonons: $\omega(q) = \omega_{LO}$ [KAK02; MZ07]. Since the free energy rotation is the same for every wave number q, it is possible to write:

$$\partial_t \langle \sum_q b_q \rangle |_{\text{free}} = -i \sum_q \omega_q \langle b_q \rangle = -i\,\omega_{LO} \langle \sum_q b_q \rangle, \tag{3.6}$$

defining:

$$\bar{b} := \sum_q (g_q^c - g_q^v) b_q. \tag{3.7}$$

Now, only the number of phonon operators is of importance and the set of equations is simplified for matters of clarity. In the one electron and two level assumption, only three types of fermionic states are possible: G is the ground, E the excited state and T the transition from ground to excited state. All three of them can be photon(p, s)- and phonon(m, n)-assisted with ($p, s, m,$ and n integers):

$$G_{m,n}^{p,s} := a_v^\dagger a_v c^{\dagger p} c^s \bar{b}^{\dagger m} \bar{b}^n, \tag{3.8}$$
$$E_{m,n}^{p,s} := a_c^\dagger a_c c^{\dagger p} c^s \bar{b}^{\dagger m} \bar{b}^n, \tag{3.9}$$
$$T_{m,n}^{p,s} := a_v^\dagger a_c c^{\dagger p} c^s \bar{b}^{\dagger m} \bar{b}^n. \tag{3.10}$$

The Heisenberg equation of motion leads to a general set of equations of motion for different orders in electron-photon and electron-phonon coupling elements by raising or lowering the indices p, s, m, n. An observable of interest to calculate the absorption is the microscopic polarization. The dynamics are calculated via

$$\partial_t \langle T_{m,n}^{p,s} \rangle = \tag{3.11}$$
$$= -i\left[\omega_{cv} - (p-s)\omega_0 - (m-n)\omega_{LO} - i(p+s)\kappa - i\gamma\right] \langle T_{m,n}^{p,s} \rangle$$
$$- ip\, M\langle E_{m,n}^{p-1,s} \rangle - iM(\langle E_{m,n}^{p,s+1} \rangle - \langle G_{m,n}^{p,s+1} \rangle) - i\Omega(t)\left(\langle E_{m,n}^{p,s} \rangle - \langle G_{m,n}^{p,s} \rangle\right)$$
$$- i\,\langle T_{m,n+1}^{p,s} \rangle - i\,\langle T_{m+1,n}^{p,s} \rangle + i\,m\,g_v\,\langle T_{m-1,n}^{p,s} \rangle - i\,n\,g_c\,\langle T_{m,n-1}^{p,s} \rangle,$$

where $g_i = \sum_q g_q^i (g_q^{c*} - g_q^{v*})$ for $i = v, c$. A phenomenological pure dephasing γ is included and takes into account, e.g. longitudinal acoustical (LA) phonon interaction. A cavity loss considers a finite cavity photon life time: κ. The microscopic polarization is driven via spontaneous emission of photons / phonons, induced emission and absorption of photons and modulated by LO-phonon assisted higher-order polarization terms. The classical pump field is time-dependent, and can also be assumed as probe to calculate absorption spectra in the linear optics regime, cf. Sec. 3.3. The photon- and phonon assisted ground and excited state densities are source terms of the dynamics. The equation of the photon- and phonon-assisted ground state reads:

$$\partial_t \langle G_{m,n}^{p,s} \rangle = \tag{3.12}$$
$$= i\left[(m-n)\omega_{LO} + (p-s)\omega_0 + i(p+s)\kappa\right] \langle G_{m,n}^{p,s} \rangle$$
$$+ i\,M\langle T_{m,n}^{p+1,s} \rangle - i\,M\langle T_{n,m}^{s+1,p} \rangle^* + i\,s\,M\langle T_{m,n}^{p,s-1} \rangle - i\,p\,M\langle T_{n,m}^{s,p-1} \rangle^*$$
$$+ i\,\Omega(t)(\langle T_{m,n}^{p,s} \rangle - \langle T_{n,m}^{s,p} \rangle^*) + i\,m\,g_v\,\langle G_{m-1,n}^{p,s} \rangle - i\,n\,g_v^*\,\langle G_{m,n-1}^{p,s} \rangle$$

In the same way, spontaneous emission and absorption processes of LO-phonons are included. The dynamics of the excited state is given by:

$$\partial_t \langle E_{m,n}^{p,s} \rangle = \tag{3.13}$$
$$= i\left[(m-n)\omega_{LO} + (p-s)\omega_0 + i(p+s)\kappa\right] \langle E_{m,n}^{p,s} \rangle$$
$$- i\,M\langle T_{m,n}^{p+1,s} \rangle + i\,M\langle T_{n,m}^{s+1,p} \rangle^*$$
$$- i\,\Omega(t)(\langle T_{m,n}^{p,s} \rangle - \langle T_{n,m}^{s,p} \rangle^*) + i\,m\,g_c\,\langle E_{m-1,n}^{p,s} \rangle - i\,n\,g_c^*\,\langle E_{m,n-1}^{p,s} \rangle.$$

The excited state and ground state density differ strongly: a typical quantum optical feature [CSS75; SZ97]. The ground state couples to the higher and lower photon number state, but the excited state only to the higher. This asymmetry is lost in the cluster expansion, where the factorized excited and ground state density are determined via the same equations, only with a reversed sign $\partial_t \langle a_c^\dagger a_c c^\dagger c \rangle^c = -\partial_t \langle a_v^\dagger a_v c^\dagger c \rangle^c$, cf. Eq. (2.15). In this context, it is worth noting, that purely photonic $\langle c^{\dagger p} c^p \rangle$ or purely phononic expectation values $\langle \bar{b}^{\dagger p} \bar{b}^p \rangle$ do not appear in Eq. (3.11) - (3.13), indicating the full quantum

3 Quantum dot cavity quantum electrodynamics

correlation of the combined electron, LO-phonon and photon dynamics.

Since this set of equations is not closed in terms of n, m, p, s, many orders have to be calculated to reach convergence. The number of orders depend on the initial conditions, as well as the strength of the implemented parameters such as the coupling strengths of $g_{v/c}$ and M. However, this set of equations of motion models the exact solution numerically to an arbitrary accuracy, cf. Fig. 3.3.

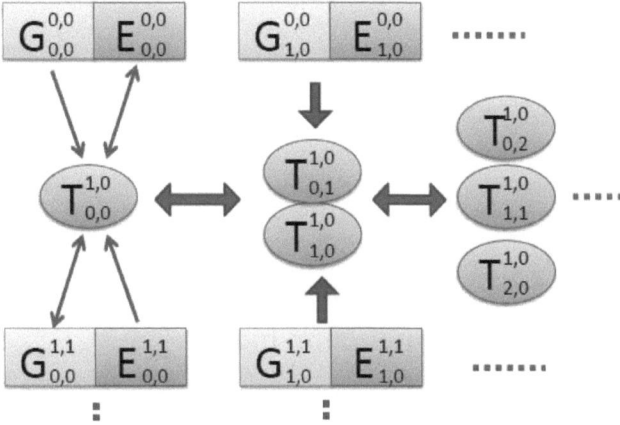

Figure 3.3: Scheme of general equations of motion within the induction method without classical pumping.

3.1.3 Initial conditions

A wide range of initial conditions can be chosen to calculate the system dynamics. The expectation values at $t_0 = 0$ are determined by the following system parameters: the photon statistics and the mean number of photons in the cavity, the temperature of the LO-phonon bath and the initial electronic state. In general, the initial conditions for a cavity field prepared in a Fock or thermal state read:

$$\begin{aligned}
\langle E^{p,s}_{n,m}\rangle(t_0) &= \rho_0^E\, \rho_p^{pt}\, \rho_n^{ph}\, \delta_{m,n}\, \delta_{p,s} \ , \\
\langle G^{p,s}_{n,m}\rangle(t_0) &= \rho_0^G\, \rho_p^{pt}\, \rho_n^{ph}\, \delta_{m,n}\, \delta_{p,s} \ , \\
\langle T^{p,s}_{n,m}\rangle(t_0) &= \rho_0^T\, \rho_p^{pt}\, \rho_n^{ph}\, \delta_{m,n}\, \delta_{p,s} \ ,
\end{aligned} \qquad (3.14)$$

with $\rho_0^E := \langle E^{0,0}_{0,0}\rangle(t_0) = \langle a_c^\dagger a_c\rangle(t_0)$, $\rho_0^G := \langle G^{0,0}_{0,0}\rangle(t_0) = \langle a_v^\dagger a_v\rangle(t_0)$ and $\rho_0^T := \langle T^{0,0}_{0,0}\rangle(0) = \langle a_v^\dagger a_c\rangle(t_0)$ denoting the probability to find the electron in the excited E or in the ground state G at time $t_0 = 0$ or with an initially assumed transition probability T.

The photon statistics enters in $\rho_p^{pt} := \langle c^{\dagger p} c^p\rangle$. The expectation value for $p > 1$ differs for coherent, thermal and non-classical states strongly [Lou83; CKR09]. Different initial conditions in dependence on the chosen photon-statistics are discussed in Sec. 3.2.3.

Initially, the LO-phonon bath is in thermal equilibrium. Using Wick's theorem [Mah90] and including the q-sum of the different wave numbers of the LO-phonon operators, arbitrary orders in the pure phonon expectation value read:

$$\rho_n^{ph} := \langle \bar{b}^{\dagger m} \bar{b}^n \rangle \delta_{mn} = n! \left(\sum_q |g_q^c - g_q^v|^2 \right)^n \bar{n}_{LO}^n, \qquad (3.15)$$

where the mean number of phonons in the LO-phonon mode $\bar{n}_{LO}$ is determined by the Bose-Einstein distribution $\bar{n}_{LO} = [\exp(\beta\hbar\omega_{LO}) - 1]^{-1}$ and $\beta = (k_B T)^{-1}$, as long as the temperature of the bath is fixed and the bath assumption is valid.

To benchmark the induction model, corresponding initial conditions are chosen to compare the induction model results with two exact analytical solutions: for the electron-photon interaction with the Jaynes-Cummings model (JCM) solution [JC63; SK93] and for electron-phonon interaction with the independent boson model (IBM) [Mah90; KAK02].

3.2 Benchmarking the induction model

The mathematical induction approach is not an exact solution in the sense, that an analytical formula describes the full quantum kinetics of the combined electron-photon and LO-phonon dynamics. Analytical solutions, such as the Jaynes-Cummings model [JC63] or the independent Boson model [Mah90] provide insights into the system dynamics without further numerical evaluation. The mathematical induction model is a framework, that generates exact solutions up to an arbitrary accuracy, such as the generating function approach[AHK99]. To proof the accuracy of this numerical solution scheme, the JCM and IBM solution are reproduced in this section.

3.2.1 Independent boson model: LO-phonon satellite peaks

In the independent boson model, the optical absorption of a two-level system interacting with an LO-phonon bath at non-zero temperature can be calculated exactly [KAK02]. In case of LO-phonons in GaAs, the absorption spectrum shows a set of satellite peaks at positions of a multiple of the LO-phonon energy $\hbar\omega_{LO} = 36.4$ meV. The absorption spectrum is calculated with Eq. (3.11) and setting the initial value of the polarization to $\langle T_{0,0}^{0,0} \rangle(0) =: \rho_0^T = 0.5$, corresponding to an initial excitation via a δ-pulse, and the ground state density is $\langle G_{0,0}^{0,0} \rangle(0) =: \rho_0^G = 1.0$. A pure dephasing of $\gamma_p = 2.5$ ps^{-1} is assumed. Here, one can apply a phonon operator transformation, to consider an equilibrium with the ground state (starting with an electron in the valence band, by replacing $g^c \longrightarrow g^c - g^v$ and neglecting the coupling to the valence band state). An interaction with the cavity mode ω_0 is not considered. The quantum optical coupling M is set to zero. The equations of motion are reduced to the electron-phonon

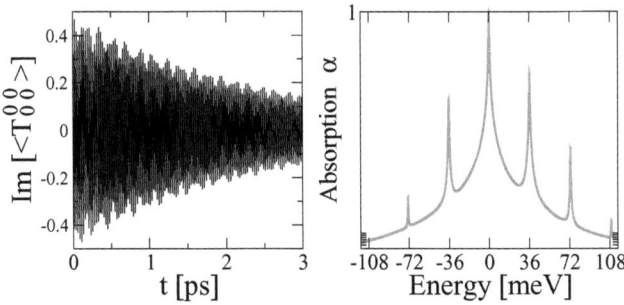

Figure 3.4: Left: Dynamics of the imaginary part of the polarization. Phonon signatures are visible in the modulated amplitude of the polarization. Right: The absorption spectrum (log scale) of the quantum dot shows LO-phonon satellite peaks at multiples of the LO-phonon energy 36.4 meV. A temperature of 300K is chosen and converging order is $m, n \geq 8$.

dynamics. By Fourier-transforming the solution of the time domain into the frequency domain, the absorption is calculated in case of a δ-pulse with [KAK02]:

$$\alpha(\omega) \propto \text{Re} \left[\frac{\langle T_{0,0}^{0,0} \rangle (\omega)}{\rho_0^T} \right]. \quad (3.16)$$

In Fig. 3.4(left), the dynamics of the imaginary part of the microscopic polarization $\langle T_{0,0}^{0,0} \rangle$ is depicted for the first 3 ps. A delta-pulse excites the system initially, and the polarization decays from its initial value due to the pure dephasing in the system, cf. Eq. (3.11). The dynamics of the polarization exhibits quantum beats with multiples of the LO-phonon frequency $\omega_{LO} = 55.3$ ps^{-1}. Here, the converging order is $m, n \geq 8$. The main resonance interferes with the satellites peaks to higher and lower energies, modulating the amplitude of the polarization during the dephasing process. The strength of the modulation depends strongly on the temperature. Here, 300 K are chosen, which leads to multiple LO-phonon satellite peaks left and right from the QD band gap frequency (shifted to zero plus polaron shift) in the absorption spectrum, cf. Fig. 3.4(right). The LO-phonon satellite peaks are broadened, since a pure dephasing is assumed. At exactly a multiple of the LO-phonon frequency, the satellite peaks appear. Experimentally, LO-phonon peaks in the absorption spectra are found in measurements, using photoluminescence excitation spectroscopy [SGB99b; HSB03; HMS+99]. The strength of the satellite peaks depends on the Huang-Rhys factor of the system. For 0 K, the peak heights of the different LO-phonon satellite peaks follow from a Poissonian distribution [MK00]:

$$p_i = e^{-F} \frac{F^i}{i!} \quad (3.17)$$

for the i-th peak with the dimensionless Huang-Rhys factor: $F = \frac{\Delta_{LO}}{\omega_{LO}}$, introducing the Polaron shift: $\Delta_{LO} = \frac{g_{\text{eff}}^2}{\omega_{LO}}$ and $g_{\text{eff}}^2 = \sum_q |g_q^c - g_q^v|^2$. This result is reproduced with the mathematical induction method.

For finite temperatures, the dependence of the satellite peaks to the zero phonon line peaks is more complicate. The mean phonon number dependence is introduced into the system, and thus induced

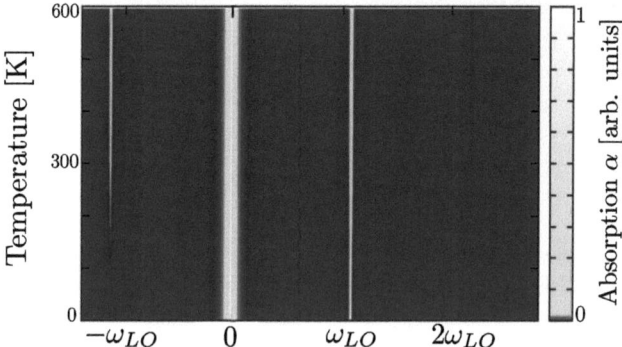

Figure 3.5: Temperature dependence of the LO-phonon satellite peaks. The quantum dot band gap frequency including polaron shift is set to zero: $\omega_{cv} - \Delta_{LO} = 0$.

emission, not only spontaneous emission of LO-phonons must be included. With increasing temperature, the IBM predicts higher satellite peaks. The induction method reproduces this result, also. In Fig. 3.5, the temperature dependence of the peak heights of the LO-phonon satellite peaks is depicted. For 0 K, Eq. (3.17) is exact for the Stokes peaks. At this temperature, only the Stokes peak appears on the higher energetic side, cf. Fig. 3.5. At elevated temperature, approximately 70 K, the first anti-Stokes peak appears in the absorption spectrum on the lower energetic side, as well as the weak second Stokes-peak. The temperature dependency is weak due to the high LO-phonon energy and thus, the small mean phonon number. Also, the effective phonon coupling determines the peak height: g_{eff}^2. In this value, the size of the quantum dot, the band gap energy and the effective mass of the electron in the valence and conduction band state as well as the shape of the QD enters. Hereby, the presented induction model reproduces the important results of the LO-phonon independent Boson model.

3.2.2 Jaynes-Cummings model (1): Vacuum Rabi splitting

Restricting the system dynamics to the electron-photon interaction, the QD acts as a two-level quantum emitter within a one mode cavity field [CTDRG89]. In a first step, the absorption spectrum is derived via Eq. (3.16). Now, the electron-phonon coupling element is set to zero ($g^{v,c} = 0$) and, again, a small pure dephasing is assumed. The system is initially excited with a δ-pulse, resulting in an initial value for the polarization of $\rho_0^T = 0.5$. No photons are inside the cavity and no further losses are included. Therefore, the system dynamics is completely determined by two coupled harmonic oscillators, for which the Jaynes-Cummings model gives the exact solution in the time domain, cf. Sec. 2.2.

Since photons and electrons are strongly coupled, new quasi-particles are formed: polaritons, giving rise to new eigen energies of the system. The derivation starts with the Jaynes-Cummings model Hamiltonian, restricting Eq. (3.2) to the electron-photon interaction, i.e. Eq. (2.2). The zero point of the energy is chosen to be in between the conduction and valence band state energy: $\omega_c = \frac{\omega_{cv}}{2}$ and $\omega_v = -\frac{\omega_{cv}}{2}$, cf. Sec 2.2. In the JCM, the dynamics is divided into subspaces of the Hilbert space, depending on the initially fixed number of photons involved in the dynamics. The two types of states are: $|\Psi_2^N\rangle = |N, c\rangle$ with N-photons in the cavity and an excited two-level system c, and $|\Psi_1^N\rangle = |N+1, v\rangle$ N+1-photons in the cavity and a two-level system in the ground state v. The Hamiltonian applied to these states leads to:

$$H|N+1, v\rangle = \hbar(\omega_0(N+1) - \omega_{cv})|N+1, v\rangle - \hbar M \sqrt{N+1}|N, c\rangle, \quad (3.18)$$

$$H|N, c\rangle = \hbar(\omega_0(N) + \omega_{cv})|N, c\rangle - \hbar M \sqrt{N+1}|N+1, v\rangle. \quad (3.19)$$

The states $|\Psi_1^N\rangle, |\Psi_2^N\rangle$ are obviously no eigenstates of the JCM-Hamiltonian. To calculate the eigenvalues, each N-part of the Hamiltonian H^N is diagonalized, defined by:

$$\left(H_{ij}^N\right) := \begin{pmatrix} \langle\Psi_1^N|H|\Psi_1^N\rangle & \langle\Psi_1^N|H|\Psi_2^N\rangle \\ \langle\Psi_2^N|H|\Psi_1^N\rangle & \langle\Psi_2^N|H|\Psi_2^N\rangle \end{pmatrix} = \hbar \begin{pmatrix} -\omega_{cv}/2 + (N+1)\omega_0 & -M\sqrt{N+1} \\ -M\sqrt{N+1} & \omega_{cv}/2 + N\omega_0 \end{pmatrix}. \quad (3.20)$$

The eigenvalues are determined by the solution of the characteristic polynomial and read:

$$\hbar\omega_{+/-}^N = \hbar\omega_0(N+\frac{1}{2}) \pm \hbar\sqrt{\frac{(\omega_{cv}-\omega_0)^2}{4} + M^2(N+1)}, \quad (3.21)$$

which results in the Jaynes-Cummings ladder with different rungs and Rabi splitting in dependence of the photon number N [WRI02; SK93; SKK08]. The vacuum Rabi splitting ($N = 0$) and in case of resonance ($\omega_{cv} = \omega_0$) has the value of $\omega_+^0 - \omega_-^0 = 2M$ and is twice the vacuum Rabi frequency. Due to the electron-photon interaction, the excited state energy splits up into a dressed state, leading to two resonance frequencies in the system. In Fig. 3.6(left), the polarization dynamics are plotted. The pure dephasing leads to a decay of the polarization. After 30 ps, the polariation vanishes. However, the electron-photon interaction leads to an additional oscillation behavior and modulates the polarization in dependence of the Rabi frequency of the system. This amplitude modulation is the signature of the Rabi splitting, depicted in Fig. 3.6(right). The QD band gap frequency is set to zero. Instead of one Lorentzian peak at $\omega = 0$, two peaks appear, marking the polariton frequencies in case of $N = 0$ photons in the cavity. From this kind of spectra, the electron-photon coupling strength can be determined [KGK+06; WNIA92; YSH+04; MKH08].

In case of a detuning δ between the QD resonance and the cavity mode, which exceeds the coupling strength $\delta \gg M$, the M-contribution in the square root in Eq. (3.21) is negligible and the eigenenergies are again the QD band gap frequency and the cavity frequency, alone. In this case, two peaks are visible in the spectrum. A large resonant peak of the QD and a small cavity peak, only weakly driven by the initial delta-pulse. In the transition regime, when the detuning $\delta \approx M$, the polariton state starts

3 Quantum Dot Cavity Quantum Electrodynamics

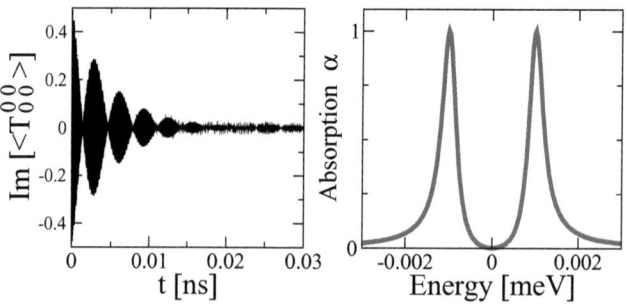

Figure 3.6: Left: Dynamics of the imaginary part of the polarization. Photon signatures are visible in the modulated amplitude of the polarization. Right: The absorption spectrum of the quantum dot shows the vacuum Rabi splitting in case of $N = 0$ photons in the cavity and resonance between the cavity mode and the QD band gap frequency.

to form, which leads close to the resonance to an anti-crossing between the cavity mode and the QD band gap frequency. In Fig. 3.7, the QD spectrum is plotted for different detunings (y-axis) of the

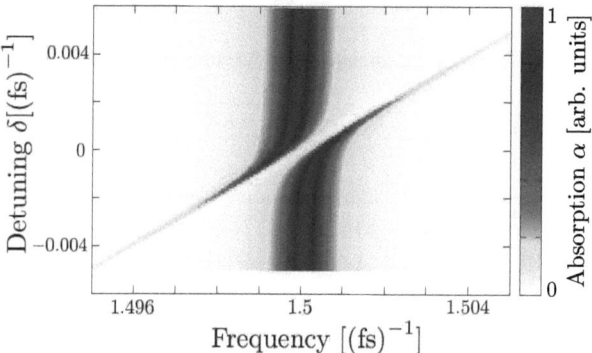

Figure 3.7: Benchmark: anti-crossing for small detunings between the cavity mode and the QD band gap frequency: pure dephasing $\gamma_p = 0.002$ fs^{-1}, coupling strength $M = 0.5$ ps^{-1} and no photons in the cavity $N = 0$. Detuning $\omega_{cv} - \omega_0 = \delta$ on y-axis, x-axis is the absorption frequency, the anti-crossing is clearly visible.

cavity mode from the QD band gap frequency (x-axis). The electron-coupling strength is chosen as $M = 0.5$ ps^{-1} and no photons are initially in the cavity: $N = 0$. If the detuning is in the order of magnitude larger than the electron-photon coupling strength, two peaks appear in the spectrum, e.g. $\delta = 5.0$ fs^{-1}. However, with decreasing detuning, the cavity peak becomes larger, until the polariton state starts to form and a splitting appears, e.g. $\delta \approx 0$. Both, the cavity frequency and the QD band frequency changes. A typical anti-crossing behavior is predicted by the exact model of Jaynes and Cummings and here, calculated via the induction method visible in Fig. 3.7.

3.2.3 Jaynes-Cummings model (2): Collapse and revivals

In the JCM, also the time dynamics of the observables are calculated exactly and additionally, different photon-statistics can be computed via a superposition of the general JCM solution, cf. Sec 2.2. Hence, another benchmark for the induction model is the reproduction of these photon-statistics induced dynamics, such as the well-known phenomenon of collapse and revivals in case of an initially coherent superposition of photon states in the cavity [ENSM80; RWK87; NSME81]. The system is closed, as assumed in the JCM. No dissipative processes, e.g. cavity losses or pure dephasing are considered: $\kappa = \gamma_p = 0\,\mu eV$. The QD is at $t_0 = 0$ in the excited state ($\langle E^{0,0}_{0,0}\rangle(t_0) = 1.0$). In dependence on the initial preparation of the cavity field, the Rabi oscillations differ strongly. The photon field can be prepared in the Fock state (a), thermal state (b) or in the coherent state (c).
The initial conditions for a photon field in the Fock state read:

$$\rho^{pt}_p = \begin{cases} \frac{N!}{(N-p)!} & \text{for } p \leq N, \\ 0 & \text{for } p > N, \end{cases} \quad (3.22)$$

where N is the number of photons in the cavity mode and p the order of the photon correlation, e.g. $\rho^{pt}_2 = \langle c^\dagger c^\dagger cc\rangle = N(N-1)$ for $p = 2$. The expectation value is zero for all p, which exceed N. In the Fock state, the mean photon number is exactly known, leading to a complete uncertainty in the phase relation of the photons. In this perspective, the Fock state is an extremely squeezed photon state and maximally non-classical [MW95]. In Fig. 3.8(a), the solution of the excited state dynamics ($\langle E^{0,0}_{0,0}\rangle(t)$) is plotted for a cavity field prepared in the Fock state. Rabi oscillations are visible with a Rabi frequency of $\Omega_N = M\sqrt{N+1}$. The quantized Rabi frequency includes the spontaneous emission of photons, which is not included in the semi-classical Rabi frequency. Even if there are no photons in the cavity, Rabi oscillations occur, as long as the electron is initially in the excited state. In Fig. 3.8(a), N is set 0 to prove this remarkable result of the JCM [JC63; SK93]. Vacuum Rabi oscillations are experimentally accessible in various experimental setups [BSKM+96; KRM+10]. The vacuum Rabi oscillation case is chosen here, to consider the strongest possible quantum correlation between the QD and the cavity mode.

If the photon field is prepared in a thermal state, the mean photon number is calculated with the Bose-Einstein statistics. The initial value is determined by Wick's theorem. Like in the case of the LO-phonons, cf. Sec. 3.1.3, the result reads:

$$\rho^{pt}_p = p!\,(N)^p, \quad (3.23)$$

with $N = [\exp(\beta\hbar\omega) - 1]^{-1}$ and $\beta = (k_B T)^{-1}$. In Fig. 3.8(b), irregular oscillations of the excited state are observable ($N = 0.1$). A pattern is not visible and not to be expected. Different Rabi frequencies interfere with each other. The induction model reproduces this complicated oscillation pattern, predicted by the JCM [SZ97; SK93]. The thermal statistics needs very high-orders in the electron-photon coupling element. The photon-correlations increase fast and high numbers exceed easily typically nu-

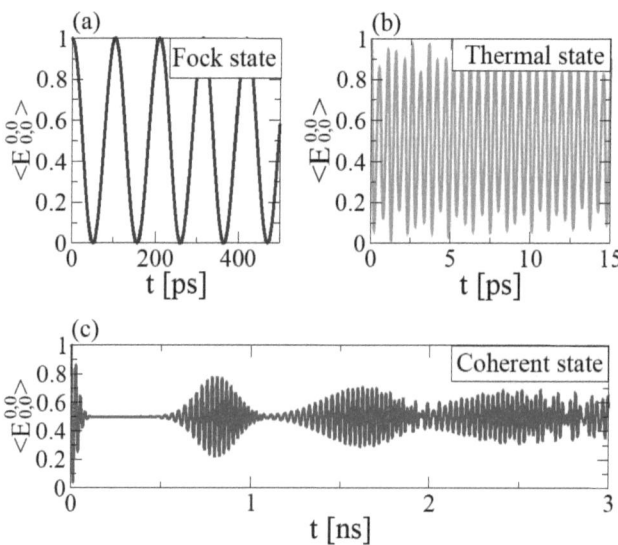

Figure 3.8: The excited state dynamics for different photon-statistics of the cavity mode: (a) Fock state $N = 0$, (b) thermal state $N = 0.4$ and (c) coherent state $N = 8$, calculated within the induction model. In case of vanishing electron-phonon interaction, the induction model reproduces the Jaynes-Cummings solution, not distinguishable in the plot (converging order: $p, s \geq 60$ for coherent and thermal states).

merical number formats. The solution, depicted in Fig. 3.8(b), takes into account orders up to $p = 60$. In principle, the solution cannot converge for very high N. Higher orders in the photon field are driven very strongly and all orders exist with a large N, cf. Eq. (3.23). However, observables such as the photon density or the $g^{(2)}(t,0)$-function, as well as the electron densities depend on these higher order correlations only very weakly due to the increasing order in the electron-photon coupling. Due to this fact, the solution converges and the limits are imposed only by the numerical implementation and the computer facilities.

Finally, the photon field can be prepared in the coherent state. Glauber-states [Gla63] are assumed. The corresponding expectation values read:

$$\rho_p^{pt} = \sum_{m,n} \frac{N^{\frac{m+n}{2}} e^{-N}}{\sqrt{m!n!}} \langle n|c^{\dagger p} c^p|m\rangle = N^p. \tag{3.24}$$

The remarkable phenomenon of periodic collapses and revivals are seen in Fig. 3.8(c) for $N = 8$ [ENSM80]. Collapses and revivals are experimentally seen in (Ref. [RWK87]). This phenomenon results from an interference between different orders in the photon-correlation and is extremely sensible to deviations from the exact solution of the JCM [Rob03]. The solution, depicted in Fig. 3.8(c), takes into account orders up to $p \geq 60$. If the dynamics is calculated for longer times $t \geq 20$ ns, higher order in p need to be calculated, since higher orders of the quantum correlations enter into the observable dynamics. However, the solution of the JCM (not distinguishable in the plot) is reproduced and proves that the induction model contains the features of a coherently-driven JCM.

The results in Fig. 3.8(a),(b) and (c) are compared with the analytic and numerical solutions of the JCM [SK93] and are found to be in complete agreement. Thus, equations (3.12) - (3.13) reproduce the JCM in the equation of motion approach, if the electron-phonon coupling is set to zero.

3.2.4 Jaynes-Cummings model (3): Analytical solution

Concluding the benchmark of the induction model regarding the electron-photon interaction, an analytical solution of the excited state density, known from the JCM, is derived to clarify the induction model without a numerical evaluation. The vacuum Rabi oscillation of the excited state density, depicted in Fig. 3.8(a), reads:

$$\langle E_{0,0}^{0,0}\rangle(t) = \cos^2(M\,t).$$

The Rabi splitting is proportional to the Rabi frequency, here the Rabi frequency is identical with the electron-light coupling element, since initially there are no photons in the cavity $N = 0$. The equations of motion, needed to derive Eq. (3.25), read with $1 = a_c^\dagger a_c + a_v^\dagger a_v$:

$$\partial_t \langle E_{0,0}^{0,0}\rangle = 2\mathrm{Im}\left[M\langle T_{0,0}^{1,0}\rangle\right] \quad (3.25)$$

$$\partial_t \langle T_{0,0}^{1,0}\rangle = -i\,M\langle E_{0,0}^{0,0}\rangle - i\,M(2\langle E_{0,0}^{1,1}\rangle - \langle c^\dagger c\rangle) \quad (3.26)$$

$$\partial_t \langle c^\dagger c\rangle = -2\mathrm{Im}\left[M\langle T_{0,0}^{1,0}\rangle\right] = -\partial_t\langle E_{0,0}^{0,0}\rangle. \quad (3.27)$$

Now, the photon-assisted excited state density is not driven in case of vacuum Rabi oscillations, and the set of equations of motion is closed. For convenience, the equations are solved for the difference between the photon denstiy and the excited state density ($W = \langle c^\dagger c\rangle - \langle a_c^\dagger a_c\rangle$):

$$\partial_t(\langle c^\dagger c\rangle - \langle E_{0,0}^{0,0}\rangle) = \partial_t W = -4\mathrm{Im}\left[M\langle T_{0,0}^{1,0}\rangle\right] \quad (3.28)$$

$$\partial_t \langle T_{0,0}^{1,0}\rangle = i\,M(\langle c^\dagger c\rangle - \langle E_{0,0}^{0,0}\rangle) = i\,M\,W \quad (3.29)$$

$$\rightarrow \partial_t^2 W = -(2\,M)^2 W. \quad (3.30)$$

This simple differential equation is solved via a cosine ansatz, bearing in mind the initial condition of $W(0) = -1$:

$$W(t) = \langle c^\dagger c\rangle(t) - \langle E_{0,0}^{0,0}\rangle(t) = -\cos(2\,M\,t) = -\cos^2(Mt) + \sin^2(Mt), \quad (3.31)$$

which gives the solution of the inversion, but also for the excited state dynamics. The initial conditions verify: $\langle E_{0,0}^{0,0}\rangle(t) = \cos^2(M\,t)$. The Rabi frequency is proportional to M, proving, that the induction model includes the JCM solution analytically. However, higher order in this truncation scheme are difficult to derive, and numerical evaluations become necessary.

3.3 LO-phonon QD cavity-QED: LO-phonon assisted cavity feeding

Cavity quantum electrodynamics (CQED) studies the interaction between a single radiation-field mode and a quantum emitter, here a semiconductor quantum dot (QD). The pursuit of QD-CQED is motivated by the possibility to tailor the quantum emission properties, e.g. the frequency and photon-statistics, within ultrasmall volumes [KGK+06; BSL+09]. Semiconductor technology paves the way for the design of monolithic structures, such as micropillars, microdisks or photonic crystal nanocavities [MKB+00; KGK+06; BSL+09]. In these systems, typical cavity-QED phenomena are demonstrated, e.g. Vacuum Rabi splitting and Purcell enhancement [LVT08], entanglement [TPT06; HPS07] and photon antibunching [MKB+00; WKU+05]. These observations show that quantum information tasks are achievable and technological applications are on the verge of realization like single and entangled photon sources or low threshold nanolasers.

Recent studies of QD-CQED show intense cavity emission, and even lasing, when the cavity mode is not in resonance with the QD. This cavity feeding is disussed widely in the literature and can be attributed to the solid-state environment of the QD [TKH+09; SDG+09]. Experimental and theoretical work confirm that longitudinal acoustical (LA) phonons play a crucial role for small detunings (few meV) and quasi-excitonic transitions for larger detunings [MKH08; Hoh10; WVT+09; TS10]. In case of large detunings, phonons are typically not taken into account as a feeding channel.

In this section, the LO-phonon interaction comes into focus and is investigated within the induction model approach to simulate the strong coupling signatures for an InAs/GaAs-quantum dot in a microcavity for different detunings, in particular for high temperatures. Strong coupling signatures become visible, if the cavity is one LO-phonon energy detuned from the QD transition energy. A clear anti-crossing is visible for the Stokes position, the anti-Stokes anti-crossing is weaker but still noticeable at 300 K.

The calculation is based on the general set of equations, cf. Eq. (3.11)-(3.13), derived by the mathematical induction, cf. Sec. 3.1 with given initial conditions. To investigate the impact of the LO-phonons, modified Rabi oscillations of the excited state density are plotted and an analytical expression for the Rabi frequency is derived. First, the initial conditions need to be discussed to set up the general set of equations of motion, adjusted to the initial conditions.

3.3.1 Initial conditions

The absorption spectrum is calculated with Eq. (3.11) by including phenomenologically a pure dephasing $\gamma = 2$ (ps)$^{-1}$. The system is excited with a delta pulse $\Omega_\delta = \Omega(t)\delta(t)$, creating an initial value of the polarization to $\langle T_{0,0}^{0,0}\rangle(0) =: \rho_0^T = 0.5$ and the ground state density is $\langle G_{0,0}^{0,0}\rangle(0) =: \rho_0^G = 1.0$. The delta pulse creates a polarization on a time scale, that the electronic system and the LO-phonon bath remains in equilibrium at $t = 0$. Material characteristics are implemented in the choice of initial conditions, here, in the calculation of the normalized effective form factor $|g_q^c - g_q^v|$ and the wave number (q) dependency, as well as the electron-photon light coupling element M. The LO-phonons are initially in equilibrium with the ground state density, which results in an operator transformation, i.e. $g^v = g^{v*} = 0$ and $g^c \longrightarrow g^c - g^v$, cf. 3.2.1. The LO-phonon bath is assumed to be initially in thermal equilibrium, where the mean number of phonons in the LO-Mode $\bar{n}_{LO}$ is given with the Bose-Einstein statistics, as long as the temperature of the bath is fixed. For $t > 0$, the LO-phonon interaction is included beyond the bath assumption, since electrons and phonons are not in equilibrium anymore and non-Markovian contributions are of interest.

3.3.2 Equations of Motion

The observable of interest to calculate the absorption is the microscopic polarization. The dynamics is calculated via

$$\begin{aligned}\partial_t \langle T^{p,s}_{m,n} \rangle = &- i[\omega_{cv} - (p-s)\omega_0 - (m-n)\omega_{LO} - i(p+s)\kappa - i\gamma]\langle T^{p,s}_{m,n}\rangle \\ &- ipM\langle E^{p-1,s}_{m,n}\rangle - iM^*(\langle E^{p,s+1}_{m,n}\rangle - \langle G^{p,s+1}_{m,n}\rangle) \\ &- i\delta^p_0 \delta^s_0 \delta^m_0 \delta^n_0 \Omega_\delta - i\langle T^{p,s}_{m,n+1}\rangle - i\langle T^{p,s}_{m+1,n}\rangle - ing^2_{\text{eff}}\langle T^{p,s}_{m,n-1}\rangle.\end{aligned} \qquad (3.32)$$

A phenomenological pure dephasing γ is included and takes into account the LA-phonon interaction. A cavity loss considers a finite photon life time: κ. The equations of the photon- and phonon-assisted ground state and excited state read:

$$\begin{aligned}\partial_t \langle G^{p,s}_{m,n}\rangle = &\; i[(m-n)\omega_{LO} + (p-s)\omega_0 + i(p+s)\kappa]\langle G^{p,s}_{m,n}\rangle \\ &+ iM\langle T^{p+1,s}_{m,n}\rangle - iM\langle T^{s+1,p}_{n,m}\rangle^* + isM\langle T^{p,s-1}_{m,n}\rangle - ipM^*\langle T^{s,p-1}_{n,m}\rangle^*,\end{aligned} \qquad (3.33)$$

$$\begin{aligned}\partial_t \langle E^{p,s}_{m,n}\rangle = &\; i[(m-n)\omega_{LO} + (p-s)\omega_0 + i(p+s)\kappa]\langle E^{p,s}_{m,n}\rangle \\ &- iM\langle T^{p+1,s}_{m,n}\rangle + iM\langle T^{s+1,p}_{n,m}\rangle^* + img^2_{\text{eff}}\langle E^{p,s}_{m-1,n}\rangle - ing^2_{\text{eff}}\langle E^{p,s}_{m,n-1}\rangle.\end{aligned} \qquad (3.34)$$

Via Eq. (3.32)-(3.34), the LO-phonon assisted cavity feeding can be calculated. Note, the ground state density does not couple to the LO-phonons due to the initial conditions and the applied operator transformation. The δ-pulse is included only in the polarization for $p = s = n = m = 0$, which corresponds to the linear excitation regime.

3.3.3 LO-phonon QD cavity-QED: Optical absorption spectrum

In Fig. 3.9, the full absorption spectrum $\alpha(\omega, \Delta)$ of the LO-phonon QD-CQED is depicted for a temperature of 300 K with ω as the Fourier transform frequency and the detuning between the QD and the cavity mode: $\Delta = \omega_{cv} - (\omega_0 - \Delta_{LO})$. As band gap frequency is 1.5 eV chosen [Sti01]. Due to the polaron shift ($\Delta_{LO} = \frac{g^2_{\text{eff}}}{\omega_{LO}}$), the main peak is shifted compared to the case without LO-phonons. For zero detuning $\Delta = 0$, the cavity mode and the band gap frequency are on resonance and an anti-crossing is visible, which depends on the light coupling strength M, cf. Sec. 3.2.2. The anti-crossing is a signature for strong coupling between the QD and the cavity mode [LVT09; VLT09]. In time domain, Rabi oscillations occur (not shown). Their frequency depends linearly on the coupling strength. The higher the Rabi frequency, the larger is the Rabi splitting of the QD peak. For minor detunings, longitudinal acoustical (LA) phonon assisted cavity feeding occurs, but is not considered in this investigation [TS10; Hoh10; SDG+09]. The mismatch between cavity mode and QD frequency is compensated

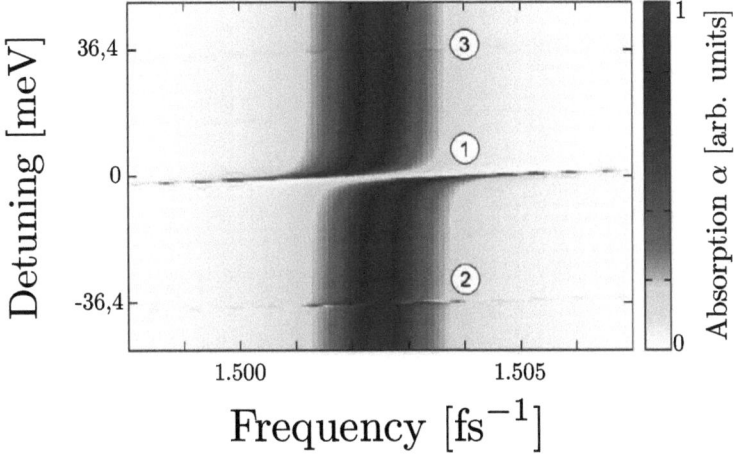

Figure 3.9: Optical absorption spectrum of an LO-phonon QD-cavity system at 300 K. Three anti-crossings are visible: (1) in case of resonance between band gap frequency and cavity mode $\Delta = 0$, which is dominated by the light coupling strength M, (2) in case of a detuning of $\Delta = -\omega_{LO}$, the Stokes contribution, which depends strongly on the effective phonon coupling strength and (3) in case of a detuning of $\Delta = \omega_{LO}$, the Anti-Stokes contribution, only visible for temperatures higher than 70 K, strongly depending on the mean phonon number.

through LA-phonon emission or absorption. This mechanism secures strong coupling features for small detunings, in order of typical longitudinal acoustical phonon frequency (few meV). It is not incorporated here, since it is focused on larger detunings, a typical longitudinal optical phonon frequency in GaAs (36, 4 meV).

If the cavity mode is detuned more than 10 meV from the band gap frequency, the Rabi splitting vanishes [Hoh10]. The spectrum is reduced to the band gap resonance peak and the cavity peak, cf. Sec. 3.2.2. But when the cavity mode matches an LO-phonon satellite peak, the anti-crossing reappears. Strongly for the Stokes peak, at a detuning of $\Delta = -\omega_{LO}$, weaker, but still visible at room temperature, the anti-Stokes contribution at $\Delta = \omega_{LO}$. The anti-crossing at a detuning position depends strongly on the material parameter, on the LO-phonon and light coupling strength. In this picture, the strong coupling signature is transported via an LO-phonon emission (Stokes contribution) or an LO-phonon absorption (anti-Stokes). Since the mean LO-phonon number at room temperature does not exceed $\bar{n}_{LO} < 0.4$ for the given LO-phonon energy in GaAs, cf. Fig. 3.14, the anti-Stokes contribution is small, in contrast to the Stokes-contribution, which is even strong at low temperatures due to spontaneous emission of LO-phonons.

For different detunings, the Rabi oscillations have a different frequency. In the Jaynes-Cummings model, which describes the main resonance without phonons, the detuning appears in a modified Rabi frequency, generalized as $\Omega = \sqrt{M^2(N+1) + \Delta^2} = \sqrt{M^2 + \Delta^2}$ for vacuum Rabi flopping $N = 0$, cf. Sec. 3.2.2 and Eq. (3.21). The detuning leads to higher Rabi frequency at the price of a reduced os-

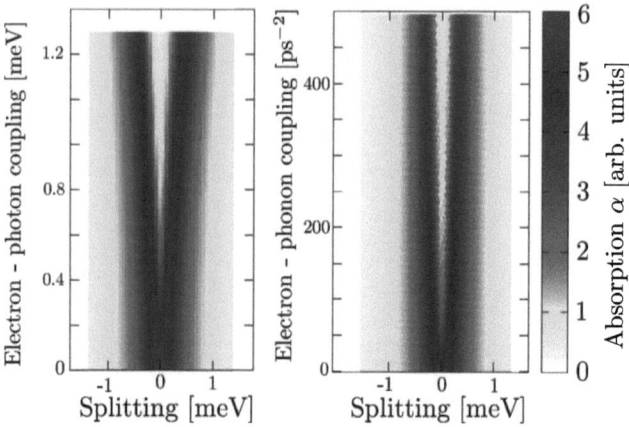

Figure 3.10: Parameter plots for a detuning of $\Delta = -\omega_{LO}$. Left: Rabi splitting for different electron-light coupling strength M (y-axis) for an effective phonon coupling of $g_\text{eff} = 75$ ps^{-2}. Right: Rabi splitting for different effective phonon coupling strengths g_eff (y-axis) and an electron-light coupling of $M = 0.75$ ps^{-1}.

cillation amplitude [SZ97]. Here, the LO-phonon interaction introduces other resonance frequencies in the system and the Rabi frequency cannot be described with the generalized formula given in the JCM.

By changing the numerical parameter, the dependence of the Rabi splitting on the coupling strength can be studied. In Fig. 3.10(left), the dependence of the Rabi splitting on the electron-photon strength M is depicted for a fixed electron-phonon coupling strength. The cavity and the QD are one LO-phonon energy detuned: $\Delta = -\omega_{LO}$ and an effective electron-phonon coupling strength of $g_\text{eff} = 75$ ps^{-2} is chosen and in agreement with experimental data [HBG+01]. For a vanishing coupling element M, the splitting also vanishes, since no interaction can take place. For a coupling element M, which is larger than $\frac{1}{2}\gamma_p$, the splitting appears. For further increment of M, the splitting increases also. First linearly, predicted by the JCM, but the deviation from the linear dependence is visible and caused by the presence of LO-phonons and LO-phonon assisted cavity feeding. In Fig. 3.10(right). The dependence of the Rabi splitting on the effective phonon-coupling strength g_eff is plotted for a electron-light coupling element of $M = 0.75$ ps^{-1}. If $g_\text{eff} = 0$, the cavity mode and the QD are spectrally too far separated and since no feeding via LO-phonon can occur, no strong coupling signature is visible. The Rabi splitting is zero. For increasing g_eff, the Rabi splitting occurs fast and in a non-linear dependence.

Now, cavity feeding via LO-phonons drives the system, but with a decreased oscillator strength due to the spectral separation. Interestingly, the Rabi splitting saturates slightly, if only the effective phonon strength of the system is increased. This indicates a complicated correlation between LO-phonon frequency, LO-phonon coupling strength, electron-photon coupling and detuning, as well as the temperature dependent LO-phonon mean number. An approximated functional dependence is derived in the next section via the Huang-Rhys factor.

3.3.4 Phonon-induced Rabi frequency modification

Now, three cases need to be discussed, as shown in Fig. 3.9 by investigating the vacuum Rabi oscillation for given detuning and without external light field, i.e. $\Omega_\delta \equiv 0$. It is assumed that the QD is initially in the excited state, $\rho_0^E = 1$ and $\rho_0^T = \rho_0^G = 0$. The equation of motion set Eq. (3.32)-(3.34) still applies. In Fig. 3.11, the dynamics of the excited state density ($\langle E_{0,0}^{0,0}(t) \rangle$) is depicted for different

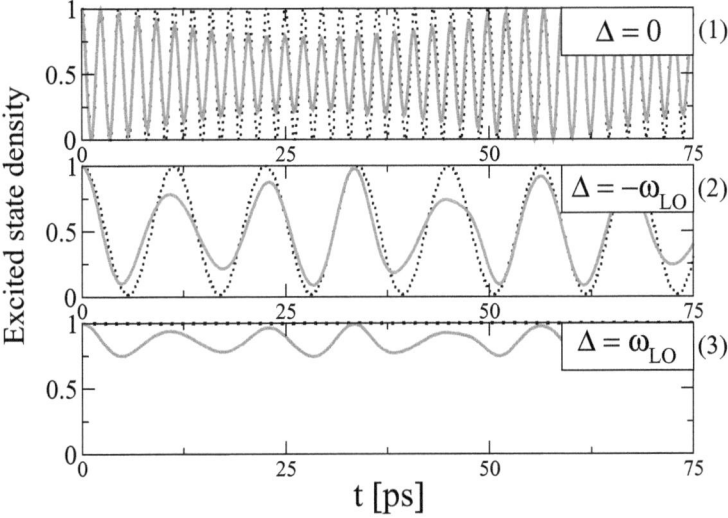

Figure 3.11: Dynamics of the excited state density for different detunings at 3 K (black, dotted line) and at 300 K (orange, solid line). (1) shows the Rabi oscillation for zero detuning, which amplitude is modulated at 300 K due to the presence of LO-phonons. (2) depicts the Rabi oscillation for a detuning of one LO-phonon frequency below the QD frequency. The Rabi oscillation frequency is strongly modified and anharmonic oscillation behavior is visible at 300 K. In case of a detuning of one LO-phonon frequency above the QD frequency (3), no Rabi oscillation is visible at 3 K. However, at 300 K LO-phonon cavity feeding takes place and a Rabi oscillation occur, with a reduced oscillation amplitude.

detunings and for 3 K (dotted, black line) and 300 K (solid, orange line). Initially, the QD is in the excited state and no photon is in the cavity.

If the cavity mode is in resonance with the QD band gap frequency (1), Rabi oscillations occur with an amplitude of 1 and a Rabi frequency equal to the JCM solution $\Omega_1 = M$. At 300 K, a beating occurs, but does not change the Rabi frequency, instead the amplitude is modulated in dependence on the effective phonon-coupling [CRCK10; AHK99]. This modulation cannot be seen in the absorption spectrum due to the strong pure dephasing. If the cavity is detuned to the Stokes peak (2), the excited state oscillates also with an amplitude of 1 but with a modified Rabi frequency Ω_2. The temperature dependence does not change the modified Rabi frequency at 300 K, which remains smaller than in the case of resonance. Remarkably, the amplitude is not decreased although the cavity mode is detuned from the band gap frequency of QD. In case of a detuning to the anti-Stokes peak (3), the temperature dependence is very important and strong. At 3 K, no oscillation occur. The cavity feeding is not active for low temperature, since the anti-Stokes process depends on the mean LO-phonon number, which is at 3 K due to the high energy of 36.4 meV negligible. Due to the lack of cavity feeding, the excited state density oscillates very fast with a negligible amplitude. However, at 300 K, cavity feeding occurs. The phonon occupation number is increased to 0.3 and an oscillation occurs with the same modified Rabi frequency $\Omega_3 = \Omega_2$, compared to the Stokes case, but with a decreased amplitude. The amplitude, but not the Rabi frequency depends on the mean phonon number.

The modified Rabi frequency due to the LO-phonon interaction depends on the oscillator strength at the given frequency. In the IBM [Mah90], the absorption spectrum of a QD interacting with an LO-phonon bath can be calculated exactly. For 0 K, the peak heights/oscillator strengths of the different LO-phonon satellite peaks follow from a Poissonian distribution in comparison to the main resonance:

$$\xi_i = e^{-F} \frac{F^i}{i!} \tag{3.35}$$

for the i-th peak with the dimensionless Huang-Rhys factor: $F = \frac{\Delta_{LO}}{\omega_{LO}}$ [MK00]. Since the splitting depends on the peak height/oscillator strength, the splitting at the Stokes (2) and anti-Stokes position (3) can be derived from the splitting ($\Omega_1 = M$) in case of resonance between QD band gap frequency and cavity mode (1) and is decreased via the Huang-Rhys factor for $i = 1$ (first peak from the main resonance):

$$\frac{\Omega_2}{\Omega_1} \approx \xi_1 = \frac{g_{\text{eff}}^2}{\omega_{LO}^2} e^{-\frac{\Delta_{LO}}{\omega_{LO}}} \longrightarrow \Omega_2 \approx \frac{M g_{\text{eff}}^2}{\omega_{LO}^2} e^{-\frac{\Delta_{LO}}{\omega_{LO}}}. \tag{3.36}$$

This expression is not exact for elevated temperatures. The modified Rabi frequencies are only proportional to the Huang-Rhys factor, since other contributions enter in the electron dynamics, e.g. the detuned Rabi frequency from the QD band gap frequency, and the temperature is 300 K. However, parameter studies, cf. Fig. 3.10, suggests an approximately linear dependence of the anti-crossing at the Stokes and anti-Stokes position on the electron-photon coupling M and the effective phonon coupling g_{eff}^2. The modulation of the amplitude shows a more complex behavior: anharmonicities appear, cf. Fig. 3.11(2) and (3). More frequencies enter the system dynamics at 300 K due to the LO-phonon induced beating.

3.4 LO-phonon QD cavity-QED: Degradation of photon-statistics

Beyond spectral properties of the LO-phonon QD cavity-QED system, discussed in the last section, with inherent cavity feeding due to the presence of LO-phonons, the time-dynamics include other observables of interest. Next to the polarization, observables such as the photon density or the normalized intensity-intensity correlation function $g^{(2)}(t,0)$-function are affected strongly by the semiconductor environment [UGA+07; MIM+00; AVB+09]. In general, the semiconductor is assumed to destroy quantum correlation via enhanced dephasing [HPS07]. In this section, the induction model is applied to the LO-phonon QD cavity-QED and demonstrates an enhancement of quantum correlation, induced by phonons, leading to an LO-phonon induced anti-bunching of the cavity field [CRCK10]. Photon-photon correlations are in focus of current research to exploit intrinsic features of the quantized light field, such as photon-bunching ($g^{(2)}(t,0)>1$) for measurement techniques [VSDS05; BO06; SBS06], coherence properties of laser emission above threshold ($g^{(2)}(t,0)=1$) [Wie09; UGA+07; AVB+09], or anti-bunching ($g^{(2)}(t,0)<1$), e.g. single-photon emission ($g^{(2)}(t,0)=0$) to control quantum information processes, quantum cryptography protocols or measurement techniques [YKS+02; BUA+05; BSPY00; LST+09]. Since QDs are grown and controlled in their optical properties [KGK+06; YSH+04] via growth techniques, strain [SWL+09], lateral fields [VUH+07], etc., quantum emitter sources based on semiconductor QDs are of great interest. To tailor the quantum light emission from QDs, embedded in a semiconductor environment, many-particle contributions, such as LO-phonons need to be taken into account. In particular in the single-photon limit, in which the fluctuation around the mean value is in the same order of magnitude of the mean number itself, quantum correlations are strongly modulated by for example the LO-phonon interaction. The single-photon limit has gained great importance recently. Unfortunately, quantum correlations are highly sensible to factorization approaches, but at the same time, those correlations are feasible for technological application, such as entanglement or anti-bunching.

To simulate and investigate the complex system dynamics on a microscopical level, quantum correlations must be treated as exact as possible. Therefore, the induction model is applied to the single-photon limit to demonstrate the strength of this approach, to treat two interactions simultaneously up to an arbitrary accuracy, including non-Markovian effects and the impact of LO-phonons on the $g^{(2)}(t,0)$-function and the photon density. Hereby, the induction model as a theoretical framework reveals possibilities to control the quantum light emission via external parameters such as the temperature and proves the importance of non-Markovian, non-factorized theoretical frameworks in the search for advantageous properties of semiconductor quantum optics.

3.4.1 Initial conditions and equations of motion

At the starting point, the QD is brought via a short pulse in the excited state on a time scale that the LO-phonon bath remains at $t=0$ in equilibrium: $\rho_0^E = 1$. The cavity is prepared either in a Fock or thermal light state. The initial conditions are calculated via the method described in 3.1.3 with Eq. (3.14). The LO-phonon bath is assumed to be initially in equilibrium and the mean number of phonons in the LO-Mode $\bar{n}_{\omega_{LO}}$ is given with the Bose-Einstein statistics. Two situations are investigated: 3 K and 300 K. No external pumping is assumed, as well as for investigation purposes, losses are not included: $\kappa = 0$ and $\gamma_p = 0$. The cavity mode (ω_0) and the QD transition (ω_{cv}) are in resonance with respective to the polaronshift $\omega_{cv} = \omega_0 + \Delta_{LO}$. The LO-phonon bath is initially in equilibrium with the ground state, cf. Sec. 3.3. In this case, the set of equations of motion Eq. (3.11)-(3.13) are simplified. The QD transition dynamics read:

$$\begin{aligned}\partial_t \langle T_{m,n}^{p+1,p}\rangle =\;& i(m-n)\omega_{LO}\langle T_{m,n}^{p+1,p}\rangle - i(p+1) M\langle E_{m,n}^{p,p}\rangle - ing_{\text{eff}} \langle T_{m,n-1}^{p+1,p}\rangle \\ & - iM(\langle E_{m,n}^{p+1,p+1}\rangle - \langle G_{m,n}^{p+1,p+1}\rangle) - i\langle T_{m,n+1}^{p+1,p}\rangle - i\langle T_{m+1,n}^{p+1,p}\rangle.\end{aligned} \quad (3.37)$$

Due to the quantum optical dynamics, without external pumping, the annihilation and creation operators of the photon field remain in constant relation. Only transitions are driven with one more or one less annihilation operator than the creation operator. In comparison to Eq. (3.11), p, s is replaced by only p. Excited state and ground state dynamics have always the same number of creation and annihilation operators. Non-diagonal contribution are not driven in this setup.

$$\begin{aligned}\partial_t \langle G_{m,n}^{p,p}\rangle &= i(m-n)\omega_{LO}\langle G_{m,n}^{p,p}\rangle - 2\text{Im}\left[M\langle T_{m,n}^{p+1,p}\rangle + p\, M\langle T_{m,n}^{p,p-1}\rangle\right] \\ \partial_t \langle E_{m,n}^{p,p}\rangle &= i(m-n)\omega_{LO}\langle E_{m,n}^{p,p}\rangle + 2\text{Im}\left[M\langle T_{m,n}^{p+1,p}\rangle\right] + ig_{\text{eff}}(m\langle E_{m-1,n}^{p,p}\rangle - n\langle E_{m,n-1}^{p,p}\rangle).\end{aligned} \quad (3.38)$$

With Eq. (3.37)-(3.38), the combined, and fully correlated electron, photon and phonon dynamics can be evaluated for different photon-statistics and phonon initial values, e.g. the temperature. The results reported here apply also to other materials, as long as the Rabi frequency $\Omega_N = M\sqrt{N+1}$ with N photons in the cavity is in the order of magnitude of the electron-phonon coupling strength $g_q^{(c,v)}$, i.e. the strength of the electron-phonon interaction is comparable with the strength of the electron-photon interaction:

$$\frac{\Omega_N}{|g_q^c - g_q^v|} \leq 100. \quad (3.39)$$

3.4.2 LO-phonon induced anti-bunching

Since the induction model evaluates the set of equations up to an arbitrary accuracy, it is now possible to study the impact of the LO-phonons on the quantum optical properties of the cavity mode, such as the photon-statistics. The interplay between the phonon- and photon interaction via the electronic system is most interesting in the regime, in which neither of them is dominant, i.e. the Rabi frequency Ω_N and the effective coupling strength $g_{\text{eff}} := \sqrt{\sum_q |g_q^c - g_q^v|^2}$ are in the same order of magnitude. In that case, none of the interactions is negligible and surprising physics occur. As a first example, it is shown that the electron-phonon interaction transforms a classical light state, such as a thermal light state, into a non-classical one [CRCK10].

Starting with a cavity prepared in a thermal state with initial conditions, cf. Sec. 3.2.3, and a mean photon number of $N = 0.25$. A thermal field at a frequency of the typical transition frequency of an InAs/GaAs QD can hardly reach this mean photon number value at reasonable temperatures. The initialization of the cavity mode can be intense tunable narrow-band thermal sources, or thermal light beams used for imaging [BH84; GBBL04]. Thus, the cavity mode is externally pumped and prepared artificially in a thermal state of $N = 0.25$. Higher values for the mean photon number are possible, but results in a very strong and photon dominated dynamics, for which the LO-phonon interaction becomes negligible. Here, the single-photon limit is investigated. The mean photon number for the thermal light field is chosen, that three or more photons have a negligible probability. In Fig. 3.12, the three-photon probability (p_3) is plotted for different mean photon numbers of the thermal light field, calculated via $p_n = N^n/(N+1)^{n+1}$ [SZ97]. Clearly, to limit the dynamics in the single-photon limit, i.e. $p_3 \ll 1\%$, the

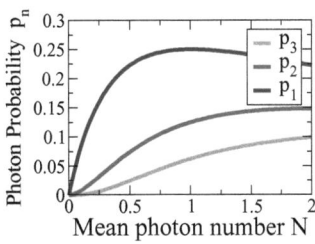

Figure 3.12: Thermal light distribution.

mean photon number must be clearly below 0.3. The thermal mixed state probability increases fast for higher mean photon numbers. Depicted in Fig. 3.12, a higher mean photon number $N > 0.6$ results in three-photon probabilities of about 5%, which are not negligible and beyond the single-photon limit.

In Fig. 3.13(a), the dynamics of the photon density is plotted for different temperatures. The QD is initially in the excited state and the photon density starts with $N = 0.25$ and increases fast, since in the spontaneous emission process a Fock photon is created and the QD electron relaxes into the ground state. The maximum value of the photon density is $N = 1.25$. Due to the strong coupling between the cavity mode and the QD electron, the photon is absorbed again. The thermal light field introduces

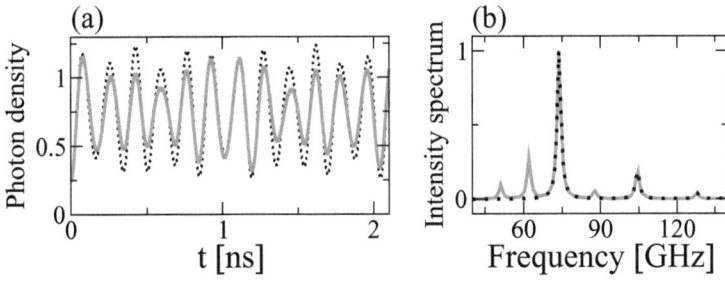

Figure 3.13: (a) Photon density of a cavity mode in the thermal state for two temperatures of the LO-phonon bath. At 3 K (black, dotted line), the photon density fluctuates strongly. At 300 K (orange, solid line), the fluctuation is decreased. (b) Fourier transform of the photon density time trace. LO-phonon peaks appear at elevated temperatures.

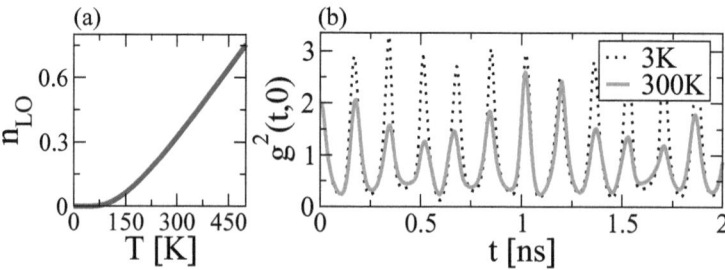

Figure 3.14: (a) Mean phonon number for different temperatures for GaAs LO-phonon energy. (b) $g^{(2)}(t,0)$ of a cavity mode in the thermal state for two temperatures of the LO-phonon bath. At 3 K (black, dotted line), $g^{(2)}(t,0)$ fluctuates strongly. At 300 K (orange, solid line), the fluctuation is decreased and leads to a lower average value of $g^{(2)}(t,0)$ from the bunching regime 1.1 at 3 K to 0.8 at 300 K.

many Rabi frequencies to the system and an irregular oscillation pattern is the result. At 3 K(dotted, black line), no LO-phonon signatures are visible. The JCM solution for a thermal light field is reproduced [SK93]. At 300 K(orangle, solid line), the oscillation pattern is changed. A modulation of the Rabi amplitude appears. This beating frequency decreases the amplitude, in general. At the right hand side, Fig. 3.13(b), the intensity spectrum is plotted, i.e. the Fourier transform of the photon density dynamics for 3 K and 300 K. At 3 K (dotted, black line), three peaks (75, 105, 130 GHz) indicate the multiple set of Rabi frequencies of the thermal light field. The higher the mean photon number N, the more peaks appear and the single peaks become more pronounced. All peaks at 3 K can be contributed to the electron-photon interaction, alone. An LO-phonon contribution is not visible. However, at 300 K (orange, solid line), additional peaks appear (around 50, 65, 90 Ghz). These peaks originate from the LO-phonon interaction and act as local oscillator, modulating the photon and electron dynamics with different temperature dependent beating frequencies. In particular, the peaks at 60 GHz and 87 GHz modulate the main oscillater frequency in between at 75 GHz. The spacing between the LO-phonon assisted peaks can be expressed via the mean LO-phonon number and the LO-phonon frequency, quantitatively $n_{LO}^2 \omega_{LO}$. For low temperatures $T < 70$ K, the mean phonon number is negligible, cf. Fig. 3.14(a). However, for elevated temperature, e.g. room temperature, the mean phonon number is around $n_{LO} = 0.3$ and the LO-phonons introduce a beating phenomenon into the system dynamics. Remarkably, this quantum beating acts as a local oscillator, comparable to a beam splitter and degradates the photon-statistics of the photon field. In contrast to the photon density in Fig. 3.13(a), on which the LO-phonon interaction has only a minor impact, the $g^{(2)}(t,0)$-function of the cavity field is strongly changed. The time trace of the $g^{(2)}(t,0)$-function is depicted in Fig. 3.14(b) for 3 K and 300 K. Initially, the cavity field is in equilibrium and prepared in a thermal state. The $g^{(2)}(t,0)$-function is equal to 2. After the dynamics starts, a Fock photon is emitted into the cavity mode and the $g^{(2)}(t,0)$-function starts to fluctuate around 2 for 3 K (dotted, black line). The

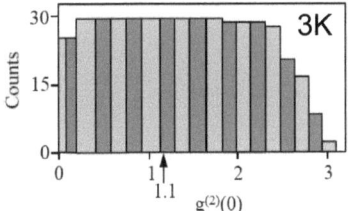 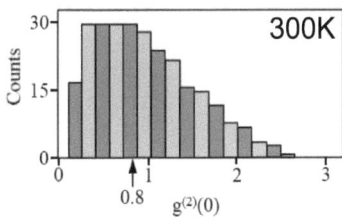

Figure 3.15: $g^{(2)}(t,0)$ of a cavity mode in the thermal state for two temperatures of the LO-phonon bath. At 3 K (black, dotted line), the photon density and the $g^{(2)}(t,0)$ fluctuates strongly. At 300 K (orange, solid line), the fluctuation is decreased and leads to a lower average value of $g^{(2)}(t,0)$ from the bunching regime 1.1 at 3 K to 0.8 anti-bunching at 300 K.

highest and lowest value of the $g^{(2)}(t,0)$-function are far apart, but the mean value stays well within the bunching regime with

$$\bar{g}^{(2)} := \frac{1}{T_a}\int_0^{T_a} dt\, g^{(2)}(t,0) = 1.1, \qquad (3.40)$$

for a fixed average time T_a, e.g. $T_a = 2$ ns. The Fock photon does not change the initial photon-statistics of the cavity mode for low temperatures. For elevated temperatures $T > 70$ K, e.g. 300 K (orange, solid line), the quantum beating leads to a modulated $g^{(2)}(t,0)$-function and surprisingly, only the highest values of the $g^{(2)}(t,0)$-function are decreased drastically. In consequence, the mean value of the $g^{(2)}(t,0)$-function drops for 300 K. The initial thermal cavity field is transformed into a non-classical one with a mean value of $\bar{g}^{(2)} = 0.8$, well within the anti-bunching regime. The LO-phonon interaction with substantial mean phonon number has a strong impact on the quantum correlation of the cavity field.

To illustrate this result, counting averages for the $g^{(2)}(t,0)$-function are plotted in Fig. 3.15 for 3 K and 300 K. It is counted, up to a given time T (2 ns), how many times the $g^{(2)}(t,0)$ function enters a given probability interval, e.g. between 0.25 and 0.3. For example, at 3 K, the $g^{(2)}(t,0)$-function reaches only two times a value larger than 3, cf. Fig. 3.14, but never at 300 K. This counting is done from the anti-bunching regime 0 to the bunching regime of 3 and plotted via bars in Fig. 3.15. The remarkable result is now clearly visible. The probability to measure a $g^{(2)}(t,0)$-value of larger than 1 is drastically decreased at 300 K. It is much more likely to measure a value indicating anti-bunching ($g^{(2)} < 1$) in comparison to the 3 K case. Noteworthy, the probability to measure extreme anti-bunching values ($g^{(2)} < 0.2$) is also decreased at 300 K. This can be explained by the degradation of the Fock statistics. To proof, that the transformation of the thermal light field is not caused by the LO-phonon impact on the Fock photon, which is spontaneously emitted from the QD, the Fock case is now investigated. The QD is initially in the excited state and the cavity contains one Fock photon. In this preparation, the photon probability distribution p_n is restricted to two values, which are non-zero: p_1 and p_2. Either there is one or two photons in the cavity, depending on the state of the QD carrier. In Fig.

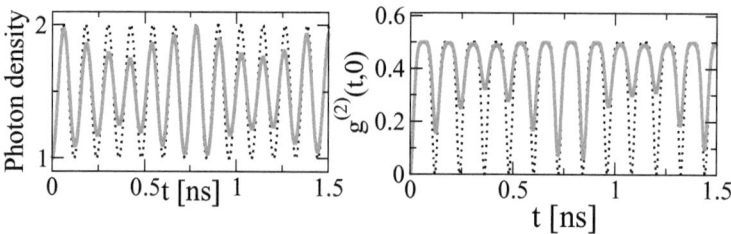

Figure 3.16: Dynamics of the photon density and $g^{(2)}(t,0)$ of a cavity mode in the Fock state for two temperatures of the LO-phonon bath. At 3 K (black, dotted line), the photon density oscillates between 1 and 2 with the Rabi frequency. At 300 K (orange, solid line), a phonon-induced quantum beating is visible, but the average value of $g^{(2)}(t,0)$ is only slightly changed and remains at approximately 0.4.

3.16, the dynamics of the photon density and the intensity-intensity correlation function $g^{(2)}(t,0)$ is plotted in case of a cavity mode prepared in the Fock state with $N = 1$ and for two temperatures. At 3 K (black, dotted line), no LO-phonon impact is visible. The photon density oscillates with the Rabi frequency $\Omega_{Rabi} = M\sqrt{2}$ between 1 and 2. At elevated temperature, e.g. 300 K (orange, solid line), the quantum beating becomes visible and the oscillation amplitude of the photon density and the $g^{(2)}(t,0)$ is decreased. If the cavity mode is prepared in a Fock state, the intensity-intensity correlation $g^{(2)}(t,0)$-function reads:

$$g^{(2)}(t,0) = 1 - \frac{1}{\langle c^\dagger c \rangle} + \frac{\langle (\Delta n)^2 \rangle}{\langle c^\dagger c \rangle^2} = 1 - \frac{1}{\langle c^\dagger c \rangle}. \quad (3.41)$$

Thus, for a system purely in the Fock state the standard deviation $\langle (\Delta n)^2 \rangle = \langle c^\dagger c c^\dagger c \rangle - \langle c^\dagger c \rangle^2$ is zero and the $g^{(2)}(t,0)$-function oscillates between 0.5 for $\langle c^\dagger c \rangle = 2$ or 0 for $\langle c^\dagger c \rangle = 1$ [MW95]. This is shown in Fig. 3.16(right) with a temperature of 3 K, at which the LO-phonon interaction is negligible (dotted, black line). In contrast to the thermal light, at 300 K the $g^{(2)}(t,0)$ function is narrowed around a higher value. Thus, the standard deviation becomes non-zero due to a phonon-induced Fock state mixing and Eq. (3.41) is not valid anymore. However, since the phonon occupation ($\bar{n}_{LO} = 0.3$) is small in comparison to the photon number (between 1 and 2), the impact on the average value of $g^{(2)}$ is small (from 0.35 to 0.39). The Fock state is only minorly transformed into a mixed state [CRCK10]. Exactly treated, the electron-LO-phonon interaction results into a phonon-induced mixing of Rabi frequencies and as a consequence into a quantum beating. Since the Fock state is a maximally squeezed photon number state, the quantum beating just leads to a mixing of different orders in the light coupling. A partial mixed state is created and it is more probable to find slightly higher $g^{(2)}$-values at elevated temperatures (300 K). However, the photon-statistics itself is not changed. This is illustrated in Fig. 3.17. The counting averages for the $g^{(2)}(t,0)$ - function are plotted for 3 K and 300 K in the Fock case. The distribution is skewed to higher $g^{(2)}(t,0)$-function values, but the mean value is only slighty increased. At 3 K (left), a broad distribution from 0 to 0.45 is visible. It is equally probable to find a $g^{(2)}(t,0)$-function in between this intervall and the mean value $\bar{g}^{(2)} = 0.35$ is well in the

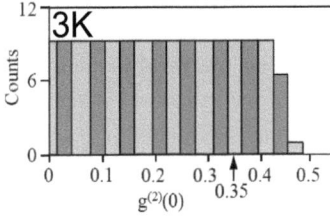 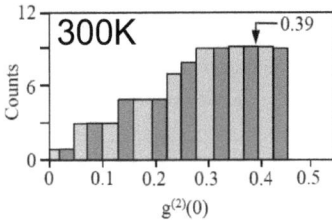

Figure 3.17: Counting averages of $g^{(2)}(t,0)$ at 3 K and 300 K in case of Fock photons in the cavity. The distribution of probable $g^{(2)}$-values is slightly skewed to higher values.

single-photon limit. At 300 K (right), it is less probable to measure a $g^{(2)}(t,0)$-function in the extreme single-photon limit $g^{(2)} < 0.3$, but the mean value remains at approximately the same value $\bar{g}^{(2)} = 0.39$. This explains, why in the thermal case the extreme single-photon values of the $g^{(2)}(t,0)$-function are also decreased at 300 K. The emitted Fock photon shifts the distribution slighty to higher values, but within the anti-bunching regime.

The LO-phonon interaction has only a minor impact on a non-classical photon distributions, such as a cavity field prepared in a Fock state, but has strong impact on classical photon distribution. As an example, a thermal cavity field was investigated. The combined, and to arbitrary accuracy evaluated electron-, photon-, and phonon dynamics result in evolution of the thermal light field into a non-classical, anti-bunched distribution. The counter-intuitive reduction of the $g^{(2)}(t,0)$-function from larger than 1 to well below 1 is enforced by a strongly modified intensity-intensity correlation $(\langle c^\dagger c^\dagger cc\rangle)$ due to the LO-phonon presence and is counter-intuitive because the photon field is made less bunched or random via interaction with a bath. The antibunching mechanism appears similar to that producing squeezed states with parametric amplification [SHY+85]. In this case, the local oscillator is the LO phonon bath Fig. 3.13, where the energetically flat phonon dispersion imparts regularity in the beating of the different Rabi oscillations. The significantly longer phonon wavelength compared to the microcavity length may be the reason for an incoherent phonon bath to have the same effect as a coherent laser local oscillator: If in a homodyne detection an incident light is mixed with a phase-matched laser local oscillator at a beam splitter, photon-antibunching can be observed [TS90]. Hereby, the strength of the induction model is proven in reproducing the Jaynes-Cummings solution, and being able to include the LO-phonon interaction in a non-Markovian, non-factorizing approach. The result occurs at high ($T_{LO} = 300K$) temperatures, and is therefore interesting for engineering applications involving nonclassical light. It is also relevant for understanding photon antibunching experiments, when blackbody radiation from the cavity wall is considered.

This model offers a wide range of possible extensions, as more electrons, more levels, even LA-phonons can be taken into account via the mathematical induction scheme and the numerical evaluation techniques. In the next section, a biexciton cascade is studied, in the weak coupling and in the strong coupling regime.

3.5 LO-phonon QD cavity-QED: Biexciton cascade

In this section, the induction model is applied to the biexciton cascade of electrons in a single QD, strongly coupled to a cavity mode. The dynamics of the biexciton cascades includes Coulomb-effects, such as energy renormalization and electron-hole exchange splitting [Tak00; Tak93; NBFZ06], and is particularly important as a possible source for polarization entangled photons [BSPY00; BUM+04]. Entangled photons constitute an experimental realization of a qubit [THM00] and are technically feasible for quantum cryptography [BBG+02], acceleration of quantum information processing [NC00] and even in measurement setups [MSO+10]. Since QDs are promising solid state sources, and can be tailored to a certain measure in their optical and electroncial properties [KKKG06], it is of great importance to understand the underlying physical processes and quantum correlations within the semiconductor environment. The induction model provides a theoretical framework to study the combined photon-, phonon- and carrier interaction with additional semi-classical electron-light interaction, or, with temperature dependent T_1 times due to LO-phonon assisted QD-carrier-WL interaction.

To simulate the biexciton cascade, first the Coulomb interaction, including the electron-hole exchange splitting, has to be considered [WSS+09]. Via an operator transformation, done by diagonalization of the Coulomb-Hamiltonian, exciton operators are introduced and the electron-photon interaction is expressed with these operators, cf. Sec. 3.5.1-3.5.2. Within the exciton basis, a general set of equations of motion is derived. As an application, the equations of motion are evaluated for the weak coupling regime and a temperature dependent degree of entanglement is calculated, cf. Sec. 3.5.5 and 3.5.6.

3.5.1 Electron-electron interaction

The two electrons in the QD interact with each other via their Coulomb potential, an additional spin-depended, repulsive exchange interaction V^{ex} is considered [WSS+09; Tak00]. In second quantization and after setting the ground energy to zero ($\hbar\omega_{v\uparrow} + \hbar\omega_{v\downarrow} + V^{vv} = 0$), the Coulomb-part of the Hamiltonian reads:

$$H|_{\text{Coulomb}} = \hbar\omega_{c\uparrow}a^\dagger_{c\uparrow}a_{c\uparrow} + \hbar\omega_{c\downarrow}a^\dagger_{c\downarrow}a_{c\downarrow} + V^{vc}(a^\dagger_{c\downarrow}a^\dagger_{v\downarrow}a_{v\downarrow}a_{c\downarrow} + a^\dagger_{c\uparrow}a^\dagger_{v\uparrow}a_{v\uparrow}a_{c\uparrow})$$
$$+ V^{cc} a^\dagger_{c\uparrow}a^\dagger_{c\downarrow}a_{c\downarrow}a_{c\uparrow} + V^{ex}(a^\dagger_{c\uparrow}a^\dagger_{v\uparrow}a_{v\downarrow}a_{c\downarrow} + a^\dagger_{c\downarrow}a^\dagger_{v\downarrow}a_{v\uparrow}a_{c\uparrow}) \quad (3.42)$$

where $\uparrow, \downarrow$ denote the spin-up, spin-down electron in ground v or excited state c. Here, the spin state of the electron $|J, m_J\rangle$ is given by the total angular momentum J and the spin projection m_j in growth direction, i.e. $|c \uparrow\rangle = |1/2, 1/2\rangle$ and $|c \downarrow\rangle = |1/2, -1/2\rangle$ for an electron in the excited state. In the ground state, the heavy hole (HH) spin state is taken into account with $|v \uparrow\rangle = |3/2, 3/2\rangle$ and $|v \downarrow\rangle = |3/2, -3/2\rangle$. Light hole and split-off spin states are energetically well below the HH and can be neglected here, cf. Fig. 3.18 [ERK+96; YC05; Sch04]. The Coulomb matrix elements V^{cc} and V^{vc} incorporating the ground state energy, which is set zero for convenience. Dark state transitions are not taken into account with $\Delta m_j \neq \pm 1$. V^{cc} is the biexciton shift, when both electrons are in the excited state. V^{vc} is the monoexciton shift, when only one electron is excited, whereas the

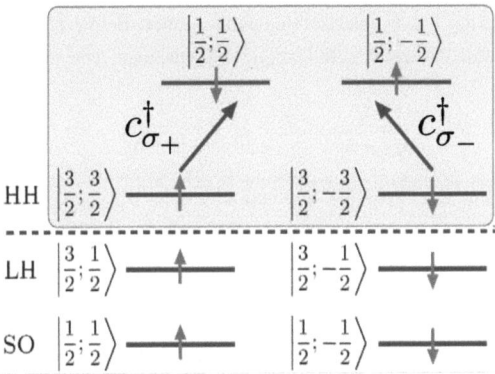

Figure 3.18: QD band structure with spin states $|J, m_j\rangle$. Light hole (LH) and split-off (SO) are energetically separated from the heavy hole state (HH), therefore the ground state spin can be taken from the HH state.

other is in the ground state [RAK+06; DAFK06; CKR09]. From the spin-orbit coupling originates an electron-hole exchange splitting (V^{ex}), in which repulsive and attraction forces induced by different spin-conformation give rise to the fine structure splitting [WSS+09; Sch04]. Additionally, it is assumed that no spin preferences exist in semiconductor QD [AKVP05], leading to the given Coulomb Hamiltonian, cf. Eq. (3.42).

Here, only four optically active states determine the system dynamics: $|v \uparrow v \downarrow\rangle, |v \uparrow c \uparrow\rangle, |v \downarrow c \downarrow\rangle$, and $|c \uparrow c \downarrow\rangle$. The dark states are neglected, such as $|v \downarrow c \uparrow\rangle$ and $|v \uparrow c \downarrow\rangle$. Four operators can be introduced for convenience and read in the two-electron basis:

$$G^{(\dagger)} = a^\dagger_{v\downarrow} a^\dagger_{v\uparrow} a_{v\uparrow} a_{v\downarrow}, \quad B = a^\dagger_{v\downarrow} a^\dagger_{v\uparrow} a_{c\uparrow} a_{c\downarrow} \quad (3.43)$$

$$P^1_1 = a^\dagger_{v\downarrow} a^\dagger_{v\uparrow} a_{v\uparrow} a_{c\uparrow}, \quad P^1_2 = a^\dagger_{v\uparrow} a^\dagger_{v\downarrow} a_{v\downarrow} a_{c\downarrow}, \quad (3.44)$$

where Eq. (3.43) are the ground state G and the biexciton state operator B, whereas in Eq. (3.44) the polarization operators $P^1_{1/2}$ are denoted. Via diagonalization, the Hamiltonian is transformed into a basis, in which the Hamiltonian with respect to the Coulomb contributions is diagonal [CKR09; RAK+06]. New eigenvalues are the result. Commuting ground and biexciton state operator (Eq. (3.43)) with the Coulomb operator show, they are already eigenvectors of the Hamiltonian. However, the polarization operators are not eigenvectors and the diagonalization lead to new exciton operators:

$$X^\dagger_m = v^m_1 P^1_1 + v^m_2 P^1_2, \quad (3.45)$$

3 Quantum Dot Cavity Quantum Electrodynamics

with $m := H/V$ replacing $1/2$ as polarization operator index. Below, H/V denotes either a horizontal or a vertical polarization. Here, the polarization is not specified. The coefficients v_i^m are given by:

$$v_1^H = -v_2^V = \frac{-\Delta_e}{\sqrt{1+\Delta_e^2}}, \quad v_2^H = v_1^V = \frac{1}{\sqrt{1+\Delta_e^2}}, \quad \Delta_e = \frac{V^{ex}}{\hbar\omega_\uparrow - \hbar\omega_H}, \tag{3.46}$$

with the new eigenvalue $\omega_{H/V}$ given by, for simplicity $V^{vc} = 0$:

$$\omega_{H/V} = \frac{\omega_{c\downarrow} + \omega_{c\uparrow}}{2} \pm \sqrt{\frac{(\omega_{c\downarrow} - \omega_{c\uparrow})^2}{4} + |V^{ex}|^2}. \tag{3.47}$$

Within the diagonalized basis, the electron operators $(G^{(\dagger)}, X_H^{(\dagger)}, X_V^{(\dagger)}, B^{(\dagger)})$ are eigenvectors of the electron part of the Hamiltonian with eigenvalues $(\hbar\omega_G = 0, \hbar\omega_H, \hbar\omega_V, \hbar\omega_B)$. The transformed Hamiltonian reads:

$$H|_{\text{Coulomb}} = \hbar\omega_H X_H^\dagger X_H + \hbar\omega_V X_V^\dagger X_V + \hbar\omega_B B^\dagger B, \tag{3.48}$$

with $X_{H/V}$ and B, the exciton and biexciton annihiliation operator, respectively. The calculation for more complex Coulomb-Hamiltonians with more contributions, or more states, and particle is straight-forward [RAK⁺06]. First, new eigenvectors are calculated and the coefficients are derived. The diagonalized Hamiltonian, e.g. Eq. (3.48), is used to rewrite any other interaction in the system, such as the electron-photon interaction.

3.5.2 Electron-photon interaction

The electron-photon interaction Hamiltonian of a semiconductor QD is usually expressed in terms of a dipole coupling between the interband polarization and the creation or annihilation of a photon. cf. Eq. (3.2). The optical and electronic properties of the QD determine the type of polarization and the wave number of the emitted quantum light. Independent of the given confinement symmetry, electron-hole exchange interaction leads to a splitting between dark and bright states and to a mixing of the dark and bright states, forming a dark and bright doublet. Emission lines, involving only pure states, are circularly polarized, whereas the mixed states result in emission lines showing linear polarization along the crystal direction [SSR⁺05; Sch04]. The emission lines with linear polarization are produced by a superposition of circularly polarized photons. In the rotating-wave approximation, this interaction reads:

$$H_{el-pt} = -\hbar \sum_k M_k \left(a_{v\uparrow}^\dagger a_{c\uparrow}^\dagger c_{k\sigma_+}^\dagger + a_{v\downarrow}^\dagger a_{c\downarrow}^\dagger c_{k\sigma_-}^\dagger \right) + h.a., \tag{3.49}$$

in which σ_+, σ_- denotes the polarization and k the wave number of the photon. The coupling element $M_{k\sigma}$ depends on the wave number, as long as it is not specified, or being transformed into an effective coupling, e.g. in a cavity. Here, a coupling strength dependence on the photon polarization is neglected

[AKVP05; HPS07]. To transform this electron-photon interaction Hamiltonian into the exciton basis, the dipole interaction is expressed in the two-electron basis by inserting a unity relation:

$$\mathbb{1}_\uparrow = a_{v\uparrow}^\dagger a_{v\uparrow} + a_{c\uparrow}^\dagger a_{c\uparrow}, \qquad \mathbb{1}_\downarrow = a_{v\downarrow}^\dagger a_{v\downarrow} + a_{c\downarrow}^\dagger a_{c\downarrow}. \tag{3.50}$$

The interaction Hamiltonian, cf. Eq. (3.49), reads after normal ordering and using the two-electron assumption, as well as Pauli's principle [RAK+06; CKR09]:

$$H_{el-pt} = -\hbar \sum_k \left(G^\dagger P_1^1 + P_2^{1\dagger} B \right) M_k c_{k\sigma_+}^\dagger + \left(G^\dagger P_2^1 + P_1^{1\dagger} B \right) M_k c_{k\sigma_-}^\dagger + h.a. \tag{3.51}$$

The basis transformation allows to express the electron-photon interaction in the new basis, in which the electron operators are eigenvectors of the electronic part of the Hamiltonian, cf. Eq. (3.45). This superposition of the exciton operators leads to a superposition of the photon operators, as well. It is convenient to define new photon operators, e.g.:

$$c_{kH}^\dagger = v_1^H c_{k\sigma_+}^\dagger + v_1^V c_{k\sigma_-}^\dagger, \qquad c_{kV}^\dagger = v_1^V c_{k\sigma_+}^\dagger - v_1^H c_{k\sigma_-}^\dagger, \tag{3.52}$$

which fulfill the commutation relation as well, since the coefficient $v_{1/2}^{H/V}$ are taken from the basis transformation. The Coulomb Hamiltonian determines the coefficients. If there is no spin-preference in the system, which would break the symmetry between the four possible transition in the two-electron case, and the exciton energies are degenerated ($\hbar\omega_{c\uparrow} = \hbar\omega_{c\downarrow}$), thus energetically separated only due to the fine structure splitting (V^{ex}), the electron-photon interaction consists of only two photon operators and reads:

$$H_{el-pt} = -\hbar \sum_k M_k \left(G^\dagger X_H + X_H^\dagger B \right) c_{kH}^\dagger + M_k \left(G^\dagger X_V - X_V^\dagger B \right) c_{kV}^\dagger + h.a. \tag{3.53}$$

with the new photon operators

$$c_{kH}^\dagger := \frac{1}{\sqrt{2}} \left(c_{k\sigma_+}^\dagger + c_{k\sigma_-}^\dagger \right), \qquad c_{kV}^\dagger := \frac{1}{\sqrt{2}} \left(c_{k\sigma_+}^\dagger - c_{k\sigma_-}^\dagger \right). \tag{3.54}$$

Furthermore, it can be assumed that the QD is placed inside a nanocavity, which supports two different modes, one for the horizontal (ω_0^H) and one for the vertical polarization (ω_0^V) of the emitted photons. The energy difference between the vertical polarized and horizontal polarized mode is in order of magnitude of μeV. Therefore, the coupling strength for both modes is the same, i.e. $M_{k_H} = M_{k_V} = M$ and the simplified interaction Hamiltonian reads finally:

$$H_{el-pt} = -\hbar M \left(G^\dagger X_H c_H^\dagger + X_H^\dagger B c_H^\dagger + G^\dagger X_V c_V^\dagger - X_V^\dagger B c_V^\dagger \right) + h.a. \tag{3.55}$$

Hereby, a cavity enhanced biexciton cascade can be studied. In Fig. 3.19, the interaction scheme restricted to the electron-photon dynamics is depicted and illustrates Eq. (3.55). Each cavity mode

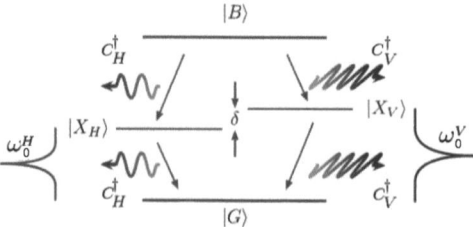

Figure 3.19: Biexciton cascade scheme. The biexciton $|B\rangle$ relaxes via the intermediate exciton states $|X_{H/V}\rangle$ to the ground state $|G\rangle$. Two photons are emitted, depending on the relaxation path. If the intermediate exciton states are energetically degenerated $\delta = \omega_H - \omega_V = 0$, the relaxation paths are indistinguishable and polarization entangled photons are produced.

supports one relaxation path via the intermediate exciton state. A cavity mode enhances only one polarization type [HHH⁺06]. The energy splitting of the intermediate exciton states depends on the excited state energy and on the exchange interaction:

$$\delta = \omega_H - \omega_V = \sqrt{|2\,V^{\text{ex}}|^2} = 2|V^{\text{ex}}|. \tag{3.56}$$

If the intermediate exciton splitting is zero, i.e. the exciton energies are the same $\omega_{c\uparrow} = \omega_{c\downarrow}$ and no fine structure splitting in the system leads to an energy shift $\delta = 0$, the four-level system reduces to a three-level system. In general, the degeneracy of the exciton states is lifted and four transition energies occur in the system: two between ground and exciton states, and two between the exciton and biexciton. In this case, the cavity modes can only be in resonance with one of this four, taken into account the polarization selection, cf. Fig. 3.19. The biexciton cascade dynamics strongly depend on how close the cavity modes and the QD transitions are in resonance, and furthermore, the degree of polarization entanglement is determined by the fine structure splitting.

3.5.3 Biexciton cascade: Equations of motion

Now, the system dynamics can be calculated within an equation of motion approach. Since the number of electrons is fixed, the induction model is applied. There are four electronic densities: the biexciton $\langle B^\dagger B\rangle$, the intermediate exciton $\langle X_H^\dagger X_H\rangle, \langle X_V^\dagger X_V\rangle$ and the ground state density $\langle G^\dagger G\rangle$, cf. Fig. 3.20. In consequence, there are four transitions: the ground state to exciton transitions $\langle G^\dagger X_H\rangle, \langle G^\dagger X_V\rangle$ and the exciton to biexciton transitions $\langle X_H^\dagger B\rangle, \langle X_V^\dagger B\rangle$. New quantities are formed in this four-level system: a ground state to biexciton transition $\langle G^\dagger B\rangle$ and the density-like quantity exciton to exciton transition $\langle X_H^\dagger X_V\rangle$, crucial for the degree of entanglement. All of these electronic states can be photon-assisted in higher order by horizontal and vertical polarized photons. Following abbreviations are convenient: $H^{m,n} = (c_H^\dagger)^m (c_H)^n$ and $V^{p,q} = (c_V^\dagger)^p (c_V)^q$. Using the Heisenberg equation of motion, the general set of

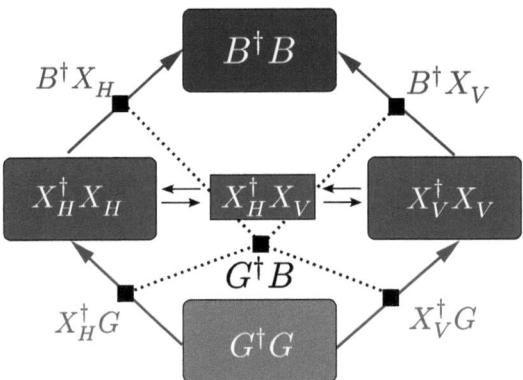

Figure 3.20: Electronic states in the biexciton dynamics. Besides to the four electronic densities: biexciton $\langle B^\dagger B \rangle$, intermediate exciton $\langle X_H^\dagger X_H \rangle, \langle X_V^\dagger X_V \rangle$ and ground state density $\langle G^\dagger G \rangle$, and the four transitions between them, the ground to biexciton state polarization $\langle G^\dagger B \rangle$ and the exciton-exciton transition $\langle X_H^\dagger X_V \rangle$ mix the horizontal $|B\rangle \to |X_H\rangle \to |G\rangle$ and vertical cascade path $|B\rangle \to |X_V\rangle \to |G\rangle$.

equations of motion is derived and proven via the induction method.
The dynamics of the biexciton density reads:

$$\partial_t \langle B^\dagger B H^{m,n} V^{p,q} \rangle \quad (3.57)$$
$$= i\left[(m-n)\omega_H^0 + (p-q)\omega_V^0 + i\kappa(m+n+p+q)\right]\langle B^\dagger B H^{m,n} V^{p,q} \rangle$$
$$- iM\langle X_H^\dagger B H^{m+1,n} V^{p,q} \rangle + iM\langle X_V^\dagger B H^{m,n} V^{p+1,q} \rangle$$
$$+ iM\langle B^\dagger X_H H^{m,n+1} V^{p,q} \rangle - iM\langle B^\dagger X_V H^{m,n} V^{p,q+1} \rangle.$$

The photon-assisted biexciton density couples to the photon-assisted polarization between the intermediate exciton and the biexciton state, cf. equations in App. 6.2. Phenomenologically, the cavity loss is introduced κ, cf. Sec. 3.1.2. Since the two-electron assumption holds, the biexciton couples only to energetic lower states, such as the intermediate exciton states via the corresponding transitions. Due to spontaneous emission processes, the exciton densities are driven stronger by the biexciton transition than by the ground state transition. The exciton-exciton transition $(X_H^\dagger X_V)$ has a significant impact on the exciton dynamics, if the two intermediate exciton levels are energetically close. The equation of motion reads:

$$\partial_t \langle X_V^\dagger X_H H^{m,n} V^{p,q} \rangle \quad (3.58)$$
$$= i\left[(m-n)\omega_H^0 + (p-q)\omega_V^0 + \omega_V - \omega_H + i\kappa(m+n+p+q)\right]\langle X_V^\dagger X_H H^{m,n} V^{p,q} \rangle$$
$$- iM\langle G^\dagger X_H H^{m,n} V^{p+1,q} \rangle + iM\, p\, \langle B^\dagger X_H H^{m,n} V^{p-1,q} \rangle + iM\langle B^\dagger X_H H^{m,n} V^{p,q+1} \rangle$$
$$+ iM\langle X_V^\dagger G H^{m,n+1} V^{p,q} \rangle + iM\, n\, \langle X_V^\dagger B H^{m,n-1} V^{p,q} \rangle + iM\langle X_V^\dagger B H^{m+1,n} V^{p,q} \rangle$$

This exchange of excitation energy is of great importance for the degree of entanglement, cf. Sec. 3.5.5. The horizontal polarized photon density can be converted completely into vertical polarized photons, if the energy splitting between the intermediate exciton levels is smaller than the cavity loss, i.e. some μeV. The conversion is driven by the ground to exciton state transitions. The ground state is populated by all relaxation processes as the energetic lowest electronic state and reads:

$$\partial_t \langle G^\dagger G H^{m,n} V^{p,q} \rangle \qquad (3.59)$$
$$= i \left[(m-n)\omega_H^0 + (p-q)\omega_V^0 + i\kappa(m+n+p+q) \right] \langle G^\dagger G H^{m,n} V^{p,q} \rangle$$
$$- iM\, m \langle X_H^\dagger G H^{m-1,n} V^{p,q} \rangle - iM \langle X_H^\dagger G H^{m,n+1} V^{p,q} \rangle - iM\, p \langle X_V^\dagger G H^{m,n} V^{p-1,q} \rangle$$
$$- iM \langle X_V^\dagger G H^{m,n} V^{p,q+1} \rangle + iM\, n \langle G^\dagger X_H H^{m,n-1} V^{p,q} \rangle + iM \langle G^\dagger X_H H^{m+1,n} V^{p,q} \rangle$$
$$+ iM\, q \langle G^\dagger X_V H^{m,n} V^{p,q-1} \rangle + iM \langle G^\dagger X_V H^{m,n} V^{p+1,q} \rangle.$$

Eq. (3.57) - (3.59) describe the cavity enhanced biexciton cascade and form a complete set with the equations in App. 6.2. More interactions can be considered straightforwardly, such as classical pumping of the biexciton density, or the LO-phonon interaction. However, this set of equations of motion contains already interesting features and can be studied in the strong and weak coupling regime.

3.5.4 Biexciton cascade: Dynamics in the strong coupling regime

To study the biexciton cascade dynamics, initial values have to be chosen, e.g. a populated biexciton density with $\langle B^\dagger B \rangle(0) = 1$ and no photons in the cavity: $\langle c_H^\dagger c_H \rangle(0) = \langle c_V^\dagger c_V \rangle(0) = 0$. The coupling strength is set to $M = 100$ μeV and a high-Q cavity is assumed $\kappa = 20$ μeV. Phonon- or Coulomb induced pure dephasing is not included, and the biexciton, and mono-exciton shift is neglected.
In Fig. 3.21(a) and (b), the dynamics of the electronic densities are plotted without dissipation processes: $\kappa = 0$. In the case of complete resonance between the excitonic levels and the cavity mode ($\omega_H = \omega_V = \omega_V^0 = \omega_H^0$), the four-level system reduces to a three level system with a degenerated intermediate exciton level with $\delta = 0$. In Fig. 3.21(a), the resonant case is plotted. The biexciton density (blue line) decays in dependence of the assumed electron-photon coupling strength and the exciton densities (red line) are built up. Both exciton densities are driven equally and are not distinguishable in this plot. With increasing exciton densities, the ground state density (green line) also builds up due to spontaneous emission processes. At 20 ps, the ground state reaches its maximum value and no excitation is in the system, but two photons with horizontal and vertical polarization have been emitted into the cavity mode. Since strong coupling is assumed, the exciton and biexciton density is driven by the cavity photons and the oscillation starts after approximately 40 ps again. In Fig. 3.21(b), a detuning between the exciton levels of 10 μeV is assumed. The same initial values are valid. Due to the detuning in the system, between cavity mode and exciton densities, the dynamics of the exciton densities differ (red and orange line). The Rabi oscillation of the biexciton and ground state is irregular. The amplitude is modulated. Eventually, the value of $\langle B^\dagger B \rangle = 1$ is reached, but after a longer time

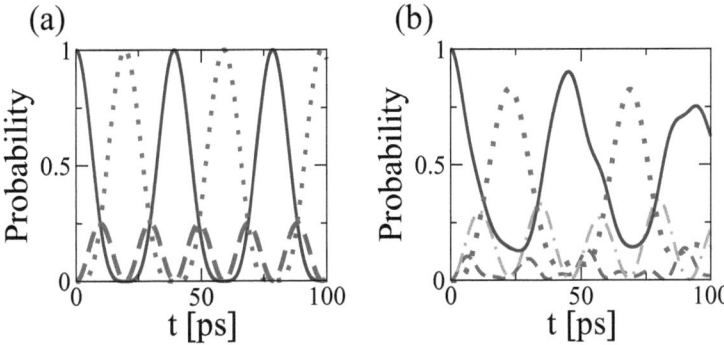

Figure 3.21: Biexciton dynamics in the strong coupling for degenerated exciton energies $\delta = 0$ (a) and for a detuning $\delta = 10\ \mu\text{eV}$. (a): The biexciton density (blue, solid line) decays and the exciton densities (dashed, red) build up and are not distinguishable, until both biexciton and exciton densities are decayed completely and the system enters the ground state state (green, dotted). (b): The exciton energies are distinguishable (red, dashed and orange, dashed-dotted) and are differently strong driven.

(not shown).
Experimentally accessible are the transmitted photon densities and intensity-intensity correlation from the cavity mode. The photon density appears in the set of equations Eq. (3.57) - (3.59) not independently, only in correlation with transitions and electron densities. This is a typical feature of a strongly coupled system, since factorization cannot be applied and the full system dynamics depends on the correlation between the electron and photon system. However, separate equations of motion of the pure photon dynamics are easily derived, using the same induction method. The photon densities for the number of horizontal and vertical polarized photons read:

$$\partial_t \langle H^{m,m} \rangle \tag{3.60}$$
$$= -2m\kappa \langle H^{m,m} \rangle - 2m\text{Im}\left[M\langle G^\dagger X_H H^{m,m-1} V^{0,0} \rangle + M\langle X_H^\dagger B H^{m,m-1} V^{0,0} \rangle \right],$$
$$\partial_t \langle V^{p,p} \rangle \tag{3.61}$$
$$= -2p\kappa \langle H^{p,p} \rangle - 2p\text{Im}\left[M\langle G^\dagger X_V H^{0,0} V^{p,p-1} \rangle - M\langle X_V^\dagger B H^{0,0} V^{p,p-1} \rangle \right].$$

For $m = p = 1$, these equations determine the photon density dynamics for horizontal or vertical polarized photons, whereas for $m = p = 2$, the intensity-intensity correlation is computed. In Fig. 3.22, the dynamics of the photon correlations is plotted. The initial values are chosen as in Fig. 3.21(b) with a fine structure splitting of $10\ \mu\text{eV}$. Horizontal and vertical cavity mode are in resonance with the ground to exciton state transition, respectively: $\omega_H = \omega_H^0 \neq \omega_V = \omega_V^0$. Initially, no photon is in the cavity and the biexciton is populated. The photon densities $\langle H^{1,1} \rangle$ (blue, dotted line) and $V^{1,1}$ (orange, solid line) start at zero and increase equally fast to 1, indicating that via the biexciton cascade one photon with horizontal and vertical polarization is generated, cf. Fig. 3.22(upper panel). Due to the small cavity loss, Rabi oscillations occur. The detuning between the intermediate exciton levels leads

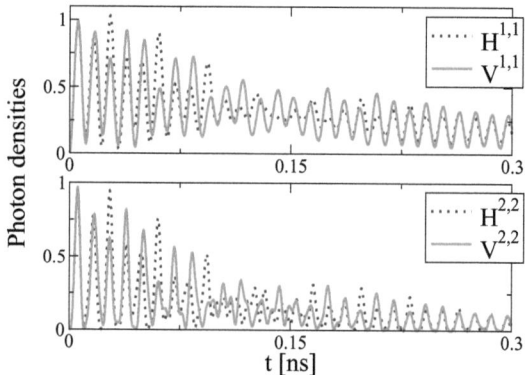

Figure 3.22: Photon density ($H^{1,1}$, $V^{1,1}$) and intensity-intensity correlation ($H^{2,2}$, $V^{2,2}$) dynamics in the biexciton cascade with a detuning of $\delta = 10$ μeV and a cavity loss of $\kappa = 10$ μeV. Horizontal and vertical driven photon quantities exhibit different oscillation patterns.

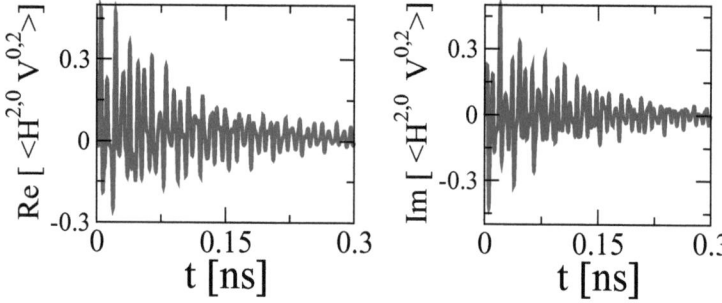

Figure 3.23: Polarization coherence dynamics ($H^{2,0} V^{0,2}$) in the biexciton cascade. This quantity determines the degree of entanglement. The higher the oscillation frequency, the smaller the degree of entanglement.

to a deviation between the oscillation behavior. The photon density in the vertical polarized mode differs from 30 ps strongly from the oscillation of the horizontal mode. The exciton-exciton transition $\langle X_V^\dagger X_H \rangle$ enforces a mixing of the excitation energy between the two cavity modes, indicated with photon densities above 1. After 150 ps, the cavity loss dominates the dynamics and the oscillations become similar again. In the lower panel of Fig. 3.22, the intensity-intensity correlation for the horizontal mode $\langle H^{2,2} \rangle$ (blue, dotted line) and the vertical mode $\langle V^{2,2} \rangle$ (orange, solid line) is plotted. The intensity-intensity correlation determines the $g^{(2)}(t, 0)$-function of the cavity mode, i.e. the photon-statistics. Again, the fine structure splitting leads to strong deviations between the horizontal and vertical mode in an irregular pattern. Since the biexciton is initially populated, it is highy unprobable that two horizontal or vertical polarized photons are generated. Therefore, the intensity-intensity correlation stays below 1 and marks a $g^{(2)}(t, 0)$-function in the anti-bunching regime.

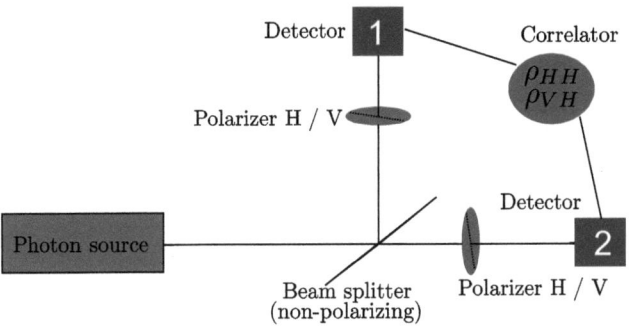

Figure 3.24: Hanbury Brown and Twiss setup. Intensity-intensity correlation in dependence on polarizers are measured. The beam splitter is non-polarizing. After correlation, the quantum state tomography elements are experimentally reconstructed, and the degree of entanglement is determined.

In terms of polarization entanglement, a coherence between the horizontal and vertical polarized photons, is of great importance. The equation of motion reads:

$$\partial_t \langle H^{m,0} V^{0,m} \rangle \tag{3.62}$$
$$= i\,m \left[(\omega_H^0 - \omega_V^0) + 2i\kappa \right] \langle H^{m,0} V^{0,m} \rangle - i\,m\,M \langle X_H^\dagger G H^{m-1,0} V^{0,m} \rangle$$
$$+ im M \left(\langle G^\dagger X_V H^{m,0} V^{0,m-1} \rangle + \langle B^\dagger X_H H^{m-1,0} V^{0,m} \rangle - \langle X_V^\dagger B H^{m,0} V^{0,m-1} \rangle \right)$$

This polarization coherence oscillates with the assumed detuning between the cavity modes, which are resonant with the ground to exciton transition of the QD system. In Fig. 3.23, the dynamics of the real (left) and imaginary part (right) of this polarization coherence is plotted. The oscillation depends on the number of photons, on the coupling strength between electrons and photons and, in particular, on the detuning between the horizontal and vertical mode. Note, the oscillation is around zero and thus, cancels time-integrated out. Here, the mean value of the real part is less than 0.002 and for the imaginary part 0.0002. But the mean value determines the degree of entanglement.

3.5.5 Biexciton cascade and entanglement

Among different proposals [EOSI04; FAT+04], very promising solid-state sources for polarization entangled photon pairs are semiconductor QDs, since a single energy level in a QD is saturated by two electrons (or holes) with opposite spins due to the exclusion principle [BSPY00]. Entanglement in its simplest form is a non-separable superposition of joint quantum states, that show non-local quantum correlations [Bel64]. The degree of entanglement can be expressed with several quantities, such as the fidelity [Jos94], the von-Neumann entropy [BDSW96] or the concurrence [Woo98]. Here, the concurrence is taken as a measure for the degree of entanglement, which is experimentally accessible via the quantum-state tomography [JKMW01; KK10]. In a Hanbury Brown and Twiss setup [HT56], intensity-intensity correlations $\langle H^{m,m} V^{p,p} \rangle$ are measured, for $m + p = 2$. Here, an experimental setup is considered, cf. Fig. 3.24, where the distance within the two light paths to the detector is appropriately adjusted to compensate the time difference between the two photon emission processes in the biexciton-cascade [SYA+06]. This enhances the probability to detect the two photons at the same time. In the quantum-state tomography [JKMW01], time-integrated measurements are done:

$$\rho_{HH} := \frac{1}{T} \int_0^T dt \, \langle H^{2,2}\rangle(t) \quad , \quad \rho_{VV} := \frac{1}{T} \int_0^T dt \, \langle V^{2,2}\rangle(t) \tag{3.63}$$

$$\rho_{HV} = (\rho_{VH})^* := \frac{1}{T} \int_0^T dt \, \langle H^{2,0} V^{0,2}\rangle(t), \tag{3.64}$$

where T is chosen large enough to consider all possible arrival times, e.g. 1 – 2 ns. The time-integrated intensity-intensity correlations ($\rho_{VV}, \rho_{HH}, \rho_{HV}$) determine the degree of the entanglement of the emitted photon pair. Neglecting other detuning sources but the fine structure splitting ($\delta \geq 0$), the biexciton decays via two relaxation paths: either $\langle B^\dagger B\rangle \to \langle X_H^\dagger X_H\rangle \to \langle G^\dagger G\rangle$ and $\langle H^{2,2}\rangle > 0$ or $\langle B^\dagger B\rangle \to \langle X_V^\dagger X_V\rangle \to \langle G^\dagger G\rangle$ and $\langle V^{2,2}\rangle > 0$. If the fine structure splitting is small ($\delta \leq 10 \, \mu$eV), the two relaxation path are not distinguishable due to the exciton-exciton transition: $\langle B^\dagger B\rangle \to \langle X_H^\dagger X_H\rangle \to \langle X_V^\dagger X_H\rangle \to \langle X_V^\dagger X_V\rangle \to \langle G^\dagger G\rangle$ and $\langle H^{2,0} V^{0,2}\rangle > 0$. As a consequence, only four combination of the intensity-intensity correlation are non-zero and the photon density matrix in the polarization sub-space ($|HH\rangle, |HV\rangle, |VH\rangle$ and $|VV\rangle$) reads:

$$\rho^{pt} := \begin{pmatrix} \rho_{HH} & 0 & 0 & \rho_{HV} \\ 0 & 0 & 0 & 0 \\ 0 & 0 & 0 & 0 \\ \rho_{VH} & 0 & 0 & \rho_{VV} \end{pmatrix}. \tag{3.65}$$

Now, the concurrence is defined via the eigenvalues λ_i (i=1,2,3,4) of the polarization sub-space density matrix with:

$$C(\rho^{pt}) := \max\{\lambda_1 - \lambda_2 - \lambda_3 - \lambda_4, 0\}, \tag{3.66}$$

where the order of the eigenvalue is defined as: $\lambda_1 \geq \lambda_2 \geq \lambda_3 \geq \lambda_4$ [SH06]. As a consequence, the concurrence has a positive value and is smaller than 1 due to the normalized character of the density matrix with Tr$[\rho^{pt}] = 1$. Here with Eq. (3.65), the concurrence can directly be calculated and expressed in dependence on the polarization coherence ρ_{HV} and reads:

$$C = 2|\rho_{HV}|. \tag{3.67}$$

For a degenerated intermediate exciton level, the horizontal and vertical decay path are not distinguishable. Therefore, the intensity-intensity correlation reads: $\rho_{HH} = \rho_{VV} = 0.5$. The polarization coherence reaches its maximum value with $\rho_{VH} = 0.5$ and the concurrence shows complete entanglement: $C = 1$.

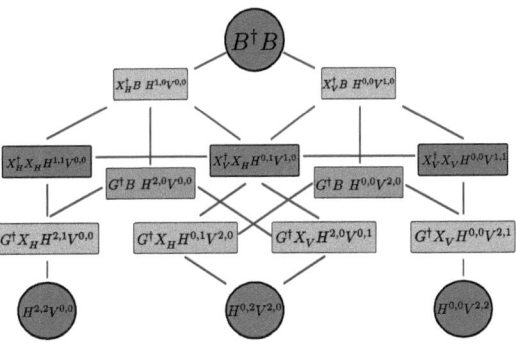

Figure 3.25: Set of equation of motion in the weak coupling regime. No induced absorption or emission processes occur. The dark blue quantities represent densities, which do not contribute to the entanglement, whereas the dark orange and red quantities directly generate a crossing of the different paths in the light orange boxes and are crucial to entanglement.

Via Eq. (3.60) - (3.62), the intensity-intensity correlations are calculated for different parameter sets. In the strong and weak coupling regime, the degree of entanglement can be discussed. In case of strong coupling, the induction method needs to be applied. An important feature, the loss of entanglement in case of a strong fine structure splitting $\delta \approx 10$ μeV, is computable also in the weak coupling regime. This is done in the next section, as the set of equations of motion reduces considerably.

3.5.6 Biexciton cascade: Weak coupling regime

In the weak coupling regime, the losses overrule the electron-photon coupling element: $M \ll \kappa$. As a consequence, the photons leave the system faster than a possible reabsorption from the electronic system may occur. Rabi oscillations are prevented and induced absorption and emission processes are negligible. The spontaneous emission dominates the system dynamics. The equations of motion are truncated to the pure cascade scheme, e.g. quantities like $\langle B^\dagger B H^{m,n} V^{p,q} \rangle$ for $m, n, p, q \neq 0$ or $\langle G^\dagger G H^{m,n} V^{p,q} \rangle$ and $\langle a_v^\dagger a_v H^{m,n} V^{p,q} \rangle$ for $m, n, p, q > 1$ vanish. E.g., the biexciton density cannot be photon-assisted and the equation of motion reads:

$$\partial_t \langle B^\dagger B \rangle = -2\Gamma \langle B^\dagger B \rangle + 2\text{Im}\left[M\langle X_H^\dagger B H^{1,0} \rangle - M\langle X_V^\dagger B V^{1,0} \rangle\right], \quad (3.68)$$

a radiative decay $\Gamma = \Gamma_{\text{rad}} = 25$ ps^{-1} into non-cavity modes is assumed and corresponds to a T_1-time, incorporated in the Weisskopf-Wigner theory [SZ97; CKR09; NBFZ06]. The cavity modes are in resonance with the ground-exciton transition: $\omega_H^0 = \omega_H$ and $\omega_V^0 = \omega_V$ with the photon life time of $\kappa = 10$ μeV. The other equations of motion, truncated to the weak coupling regime, are listed in the appendix, cf. App. 6.3.

In Fig. 3.25, the involving quantities in the cascade scheme are depicted. The color of the boxes indicate the importance of the quantity in generating a polarization entangled photon pair. The dark

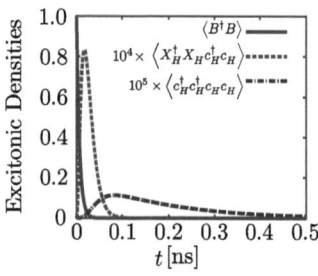 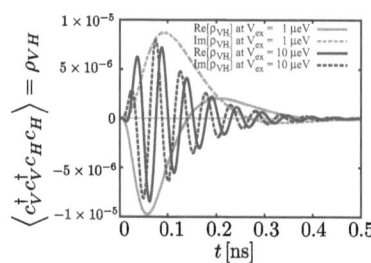

Figure 3.26: Biexciton cascade dynamics in the weak coupling regime. Left: The biexciton decays (solid line). The intermediate exciton density builds up (dashed line) and decays, while the intensity-intensity correlation is built up (dashed, dotted line). The biexciton cascade is completed. Right: The dynamics of the off polarization coherence for different fine structure splitting V_{ex}.

blue quantities represent densities, which do not contribute to the entanglement, whereas the dark orange and red quantities directly generate a crossing of the different paths in the light orange boxes and are crucial to entanglement. The dynamics start with a populated biexciton $\langle B^\dagger B \rangle$, which decays via the photon-assisted exciton-biexciton polarization. The emitted photon has either a horizontal or vertical polarization. Therefore, two decay paths are possible via the photon-density assisted intermediate exciton densities $\langle X_V^\dagger X_V H^{0,0} V^{1,1} \rangle$ and $\langle X_H^\dagger X_H H^{1,1} V^{0,0} \rangle$. However, the photon-assisted exciton-biexciton polarization couples additionally to the photon-assisted ground-biexciton state polarization $\langle G^\dagger B H^{2,0} V^{0,0} \rangle$ or $\langle G^\dagger B H^{0,0} V^{2,0} \rangle$. This characteristic quantity in a four-level system leads to a decay path mixing. This can be seen in Fig. 3.25. The ground-biexciton transition couples to ground-exciton transitions, assisted from horizontal and vertical polarized photons, such as $\langle G^\dagger X_V H^{2,0} V^{0,1} \rangle$ or $\langle G^\dagger X_H H^{0,1} V^{2,0} \rangle$. In fact, these polarization-mixed ground-exciton transitions drive the polarization coherence ρ_{HV}, which determines the degree of entanglement in Eq. (3.67).

Furthermore, the intermediate exciton states also couples to the exciton-exciton transition via an exchange of a photon with respective polarization $\langle X_V^\dagger X_H H^{0,1} V^{1,0} \rangle$. This exciton-exciton transitions drives also the polarization-mixed ground-exciton transitions. The smaller the detuning between the exciton states, the less the exciton-exciton transition oscillates and the stronger the polarization coherence is driven. But if the detuning is large $\delta \geq 10\,mu\text{eV}$, the decay path do not overlap and the biexciton decay is dominated by the emission of either two horizontal or vertical polarized photons without a polarization-mixed coherence, e.g. $\langle B^\dagger B \rangle \to \langle X_H^\dagger X_H H^{1,1} V^{0,0} \rangle \to \rho_{HH}$. In this case, the entanglement does not exist and the decay path can be reconstructed afterwards by the quantum state tomography results.

In Fig. 3.26, the biexciton cascade dynamics in the weak coupling regime is depicted. On the left, the biexciton decays (solid line), the intermediate exciton density builds up (dashed line), and the first photon is emitted. After 50 ps, the biexciton density is completely relaxed into the exciton density and due to the losses in the system, the exciton density starts to relax to the ground state via the emission of a second photon. The intensity-intensity correlation increases and reaches its maximum, when

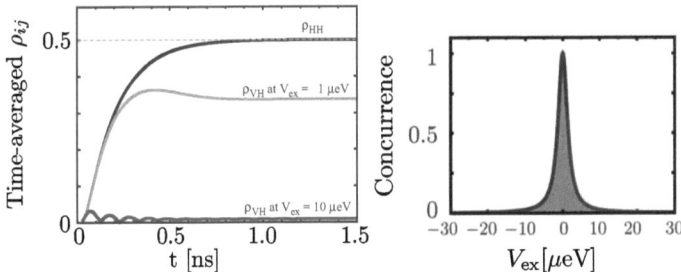

Figure 3.27: Left: Time-integrated intensity-intensity correlations. Right: The concurrence in dependence of the fine structure splitting $V_{ex} = \frac{\delta}{2}$.

the exiton density is zero. The cavity loss leads to a decay of the intensity-intensity correlation and after around 500 ps, the system dynamics stops. Two photons are emitted, either horizontal or vertical polarized or entangled. In Fig. 3.26(right), the dynamics of the polarization coherence is plotted for different detunings δ between the intermediate exciton levels. If the detuning is small $\delta = 1\ \mu eV$ (dashed line), the polarization coherence is density-like and the oscillation frequency is small. But for higher detuning values, e.g. $V_{ex} = 10\ \mu eV$, the oscillation frequency is high and the difference to the intensity-intensity correlations ρ_{HH} and ρ_{VV} is big. The oscillation frequency is determined by the detuning, cf. Eq. (3.62), of the horizontal and vertical cavity mode.

Quantum-state tomography reconstructs the polarization density matrix via averaged measurements, cf. Eq. (3.64). In Fig. 3.27(left), the time-integrated intensity-intensity correlation ρ_{HH} and polarization coherence ρ_{HV} over the time is plotted. This illustrates the importance of the detuning and the oscillation frequency of the polarization coherence. The intensity-intensity correlation is a higher order density and does not oscillate. Time-integrated saturates ρ_{HH} (blue) at the value 0.5. The values in the quantum-state tomography are normalized, that the trace of ρ^{pt} equals 1. In the biexciton cascade, ρ_{VV} and ρ_{HH} reaches time-integrated and normalized always 0.5. In contrast to the intensity-intensity correlation (orange and red line), the polarization coherence ρ_{HV} oscillates with $2\delta = 4|V_{ex}|$. If the oscillation frequency is high, compared to the duration of the biexciton cascade, time-integration leads to a vanishing value for ρ_{HV}. For a $V_{ex} = 1\ \mu eV$ (orange line), the time-integrated value of ρ_{HV} is still close to the value of ρ_{HH}. However, the oscillation frequency is around 5 ns^{-1} and the value is lowered after 0.4 ns due to the oscillation. More pronounced for a $V_{ex} = 10\ \mu eV$ (red line), the oscillation takes place on a time scale of ps and consequently the time-averaged value increases and decreases fast and remains at a low value of around 0.01. In Fig. 3.27(right), the degree of entanglement in dependence of the fine structure splitting is depicted. The cavity mode and the ground-exciton transitions are kept in resonance. If $V_{ex} \neq 0$, the cavity modes are detuned from each other: $\omega_H^0 = \omega_H \neq \omega_V^0 = \omega_V$. Consequently, the maximum entanglement is generated for a vanishing fine structure splitting. Any detuning leads to a loss of entanglement. For $V_{ex} \geq 20\ \mu eV$, the entanglement vanishes completely [CMD+10].

The fine structure splitting of the intermediate exciton states imposes a severe restriction to the generation of polarization entangled photon pairs [STS+02; SFP+02; USM+03]. Another restriction comes into play, if the QD carriers interact with the surrounding semiconductor host material, e.g. with the carriers in the wetting layer or with the bulk phonons. However, in a pure relaxation process, such as a biexciton cascade, the pure dephasing contributions may be neglegible [HPS07]. Therefore the immediate attack of occurring transitions by longitudinal acoustical and optical phonons have no impact on the degree of entanglemenent, at least in the weak coupling regime. Experiments from Hafenkramp et.al [HUM+07] show the minor impact of pure dephasing processes on the degree of entanglement. Pure dephasing increases considerably from 4 K to 30 K [KAK02], but the concurrence is only minorly decreased. Further temperature increment leads to a vanishing signal. Hafenkramp et al. contributes this temperature effect to a loss of carriers to the wetting layer. Next, this effect is taken into account via an LO-phonon coupling between WL and QD carrier states and is in agreement with the measured data. The impact of the WL - QD carrier interaction depends on the geometry, size and energy configuration of the QD. If the QD energy levels are far separated from the WL energy level, the QD is atom-like and the two-electron assumption holds [Sti01]. Yet, it may be feasible to pump the QD electrically with the ultimate goal to fabricate an entangled photon source on demand in a LED structure [SSF+10]. In this case, the WL-QD interaction cannot be neglected.

Depending on the QD size, the heavy hole state (HH) is typically separated between one and two LO phonon energies from the WL band edge, cf. Fig. 3.28(left). To connect the QD holes with the WL effectively, at least two LO-phonon processes have to be taken into account. Single phonon processes cannot contribute to the relaxation process. WL electrons are at least three LO-phonon energy separated from the QD electron state. Within the weak density regime, Coulomb interaction within the WL is negligible [DMR+10]. The microscopic relaxation rates Γ_w are derived in an effective Hamiltonian approach, based on a higher-order Markovian process. The whole two LO-phonon process is energy conserving: $\epsilon^{WL}_{vk_{res}} - \epsilon^{QD}_v = 2\hbar\omega_{LO}$, cf. Fig. 3.28(left). However, in the transition to the intermediate carrier state at ϵ^{WL}_{vk}, the energy conservation is always violated, since the hole state at **k** is always less than one LO-phonon energy separated from $\epsilon^{WL}_{vk_{res}}$. Processes, which are energy conserving, but consists of energy conservation violating subprocesses, are called higher-order Markovian process [DMR+10]. The probability amplitude of the subprocesses depend strongly on what extent the energy conservation is met.

The LO-phonon assisted relaxation rates contribute to the overall T_1-time in the system, which now reads $\Gamma = \Gamma_r + \Gamma_w$ and consists of the LO-phonon induced relaxation rate Γ_w and the radiative decay Γ_r. In contrast to Γ_r, the LO-phonon relaxation rates depend strongly on the temperature via the Bose-Einstein distribution function. For a two LO-phonon process, the temperature dependence corresponds to the temperature dependence of a Bose-Einstein distribution function to the power of two. In Fig. 3.28(right), the relaxation rate over the temperature is plotted. From a temperature of $T = 70$ K and higher, the relaxation rate exceeds the radiative decay rate $\Gamma_r = 40$ ns^{-1} and becomes the dominant damping and dephasing process in the system [CMD+10].

The temperature dependent T_1-times of the LO-phonon assisted WL carrier QD carrier scattering

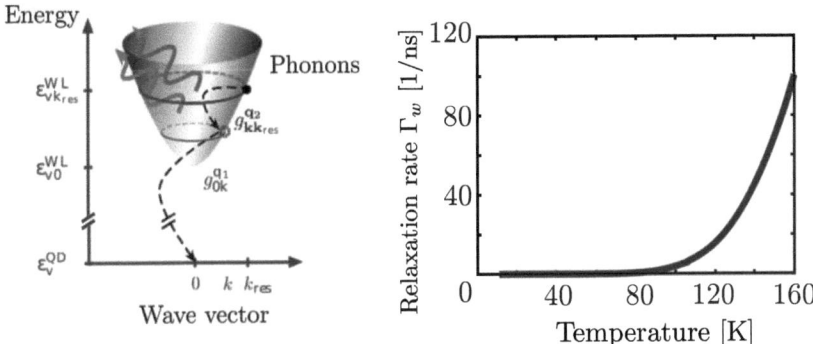

Figure 3.28: Left: LO-phonon assisted carrier relaxation in a higher-order Markovian process. Right: T_1-time in dependence on the temperature. With increasing temperature, the LO-phonon assisted scattering rate increases due to the increasing number of phonons.

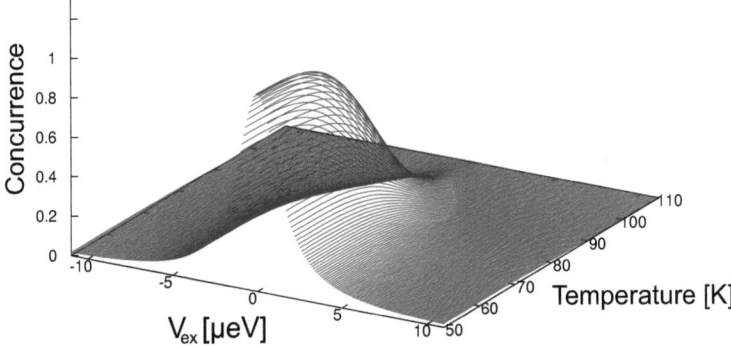

Figure 3.29: With increasing temperature, the LO-phonon assisted scattering rate increases due to the increasing number of phonons and the degree of entanglement is decreased. Until for temperatures beyond $T = 110$ K, the entanglement is entirely lost due to the WL induced damping, even for the ideal situation of degenerated exciton levels.

allow a temperature dependent analysis of the quality of entanglement of the biexciton cascade in a semiconductor QD, embedded in GaAs bulk material. In Fig. 3.29, the degree of entanglement (concurrence) is plotted for different exchange splitting strength and for different temperatures, concerning the LO-phonon damping rates. First, it is shown how the entanglement is lost with increasing V_{ex}. The FWHM of the concurrence depends on the chosen Coulomb parameters. When temperature effects of the WL states are taken into account, the concurrence is spoiled even in the ideal situation of degenerated intermediate exciton levels with $V_{ex} = 0$. For low temperatures $T < 70$ K, the concurrence remains unaffected by the WL-induced dephasing rate Γ_w, since the scattering times are well above 1 ns, thus on a larger time scale than the whole biexciton dynamics. With increasing temperature, the LO-phonon damping becomes the dominant process in the biexciton cascade. Until for temperatures beyond $T = 110$ K, the entanglement is entirely lost due to the WL induced damping, even for the ideal situation of degenerated exciton levels. Since in a pure relaxation case, i.e. a biexciton cascade in the weak coupling regime, the pure dephasing of longitudinal acoustical phonons is negligible, phonons imposes only via LO-phonon assisted scattering processes a restriction to the generation of polarization entangled photon pairs, in particular in electrically driven LED structures [SSF+10]. Depending on the host material of the QD, WL induced damping rates reduces the degree of entanglement, e.g. in GaAs at a temperature of $T = 100\ K$ in the weak coupling regime. Further investigations, in particular via the induction method in the strong coupling regime, with inherent LO-phonon coupling to the QD exciton may reveal other probabilities. E.g. the LO-phonon sidebands may enforces an erasing of the which-path information, similar to the modulation of the exciton and biexciton energies via external applied electrical fields [CG09].

4 Quantum dot - wetting layer cavity quantum electrodynamics

The fundamental model to study quantum optical properties of a two-level system interacting with a single-cavity mode is the Jaynes-Cummings model (JCM), cf. Sec. 2.2. The JCM provides an analytical solution under three conditions: (i) the one-electron assumption, i.e. expectation values such as $\langle a_1^\dagger a_2^\dagger a_3 a_4 \rangle$ are zero; (ii) only the electron-photon interaction is considered and (iii) a closed system is assumed without pumping mechanisms or losses. To place this theory into an experimental context, one may consider a stream of velocity-selected excited Rydberg atoms pass through a microwave cavity at a rate that allows only one atom to be in the cavity at any time [SK93]. Recent progress in nanotechnology provides a new fundamental scheme to study quantum optical phenomena, a semiconductor QD in a microcavity [STH⁺09; FMR⁺09].

In Chap. 3, a mathematical induction method is introduced to include the LO-phonon interaction, cavity loss and radiative decay and electrical pumping into the strongly coupled electron-photon interaction. However, a fixed number of electrons is the fundamental assumption for the induction model and two conditions for the JCM are not met: (ii) and (iii). Considering a QD in a carrier reservoir - wetting layer (WL) system, all three conditions of the JCM are not fulfilled, cf. Fig. 4.1. The model system to illustrate the proposed photon probability cluster expansion (PPCE) is described by the electron-photon Hamiltonian $H = H_0 + H_{el-pt}$ [MW95] and illustrated in Fig. 4.1(b):

$$H = \hbar\omega_0 c^\dagger c + \hbar\omega_c a_c^\dagger a_c + \hbar\omega_v a_v^\dagger a_v - \hbar M(a_v^\dagger a_c c^\dagger + a_c^\dagger a_v c). \tag{4.1}$$

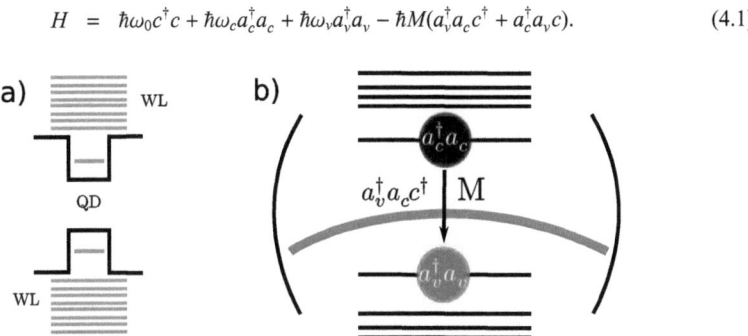

Figure 4.1: Scheme of the model system: a) a quantum dot is embedded inside a wetting layer leading to confined QD states coupled to a wetting layer continuum, b) the emission of the confined states couples to a photon mode in a cavity.

For simplicity, only one valence band state v and conduction band state c is considered with the energies $\hbar\omega_v$ and $\hbar\omega_c$, respectively. Here, the cavity mode is assumed to be in resonance with the QD band gap frequency: $\omega_0 = \omega_c - \omega_v$. The off-diagonal coupling matrix M denotes the electron-photon interaction strength. Additionally, the two level quantum dot is assumed to be embedded in an electronically occupied carrier reservoir - wetting layer. In particular, the one electron assumption is not valid and for the emission of non-classical light, many-particle correlations for the dynamics of electrons and holes confined in the QDs become important. In particular, electrical pump mechanisms are in focus of research and feasible for future technological applications, such as optical quantum-information processing or quantum cryptography [YKS+02; LO05].

For describing the quantum dynamics of semiconductor light emitters involving weak light-matter coupling and large photon numbers, an expansion involving mean field quantities and their fluctuations (often called cluster or correlation expansion) provides a well controlled theoretical scheme to treat many particle systems of electrons, phonons and photons [Fri96; AFK05; KK06; GWLJ07]. However, for semiconductor emitters operated in the limit of few photons [MIM+00] and few electronic levels, such as quantum dot-wetting layer (QD-WL) systems [GWLJ07], the cluster expansion breaks down, since it is based on the assumption that fluctuations of mean field quantities have minor influence, cf. Sec. 2.3. With the rise of quantum information and the search for proper solid state emitters, QD-WL systems for single and entangled photons are in the focus of current research [YKS+02; GWLJ07; CKR09].

In contrast to Chap. 3, it is now important to treat the strongly coupled electron-photon interaction within a many-particle perturbation approach, regarding the carrier fluctuations in a QD-WL system, in which a fixed number of electrons cannot be assumed. Therefore, there is an urgent need for the extension of the standard cluster expansion (SCE). Such an extension: the photon probability cluster expansion (PPCE) is developed in this chapter. The PPCE is a reliable approach for few photon dynamics in many body electron systems. The strength of this method is to keep the accurate results of the SCE for large photon numbers [KK06; GWLJ07], and, additionally, to include the strong coupling limit for high quality cavities [RSL+04].

This chapter is organized as following: first, the standard cluster expansion is illustrated and the Hartree-Fock factorization introduced, cf. Sec. 4.1. The intensity-intensity correlation $g^{(2)}(t, 0)$-function is computed within the SCE for a small number of photons and emitters, cf. Sec. 4.1.1, and problems are discussed. In Sec. 4.2, the PPCE is derived and PPCE and SCE solutions are compared in the single-photon limit. Within the PPCE, the environment coupling of the QD to non-lasing modes, cavity losses, pure dephasing and electrical pumping are discussed in Sec. 4.4. Finally, parameter studies of a single-QD laser in the single-photon and lasing regime are presented, cf. Sec. 4.5.

4.1 Standard cluster expansion (SCE) beyond the one-electron assumption (OEA)

A significant difference between semiconductor and atom physics is the number of carriers, which are involved in the system dynamics [Mah90]. In atom physics, the number of carriers is fixed and in most cases, only the lowest unoccupied molecule orbital (LUMO) and the highest occupied molecule orbital (HOMO) need to be considered, populated with one electron either in the excited or in the ground state. The one-electron assumption (OEA) is assumed to be valid in most of the discussed cases [Car99; SZ97]. If not the OEA, at least the ground state density is correlated to the excited state density. However, in semiconductors, the number of carriers, electrons and holes, is high and even for confined system, e.g. QDs, the OEA cannot be regarded as always valid. For shallow QD, which are energetically not far separated from the carrier reservoir of the semiconductor, the OEA is not a good approximation [BGWJ06]. More important, the number of carriers in the excited and in the ground state are not correlated, since electrons and holes have different effective masses and in case of InAs/GaAs QDs holes are energetically closer to the WL than their electrons and pumping rates differ [Sti01].

In systems without a fixed number of electrons, i.e. expectation values like $\langle a_1^\dagger a_2^\dagger a_3 a_4 \rangle$ are not vanishing, a hierarchy problem occurs. Quantities with two electronic operators $\langle a_1^\dagger a_4 \rangle$ couple to expectation values with four operators $\langle a_1^\dagger a_2^\dagger a_3 a_4 \rangle$, those to six operator expectation values $\langle a_1^\dagger a_2^\dagger a_5^\dagger a_6 a_3 a_4 \rangle$ etc [DWRK10; SCK04]. To cast the set of equations into a closed and solvable system of differential equations, the expectation values need to be factorized to eliminate the coupling to higher orders in the electronic system. The lowest order of factorization is a mean-field theory, which describes the many-particle dynamics within a single-particle interaction [Fri96].

In this section, the conventional Hartree-Fock approximation is applied to the QD cavity-QED case. The inclusion of many-particle contributions to the electron-photon dynamics leads to a modified spontaneous emission source term. However, the Hartree-Fock contributions do not solve the underlying problem of the SCE, regarding the factorization of a strongly correlated system, such as the QD cavity-QED case, cf. Sec. 2.3. However, experimental results can be explained and the calculations are valid for many emitter dynamics within the cavity, or for a high number of photons.

4.1.1 Hartree-Fock approximation within the SCE

Considering a shallow QD, with many particle contribution, the equation of motion of the polarization changes. Beyond the OEA, the microscopic photon-assisted polarization reads:

$$\partial_t \langle a_v^\dagger a_c c^\dagger \rangle^c = -iM \left(\langle a_c^\dagger a_c \rangle - \langle a_c^\dagger a_v^\dagger a_v a_c \rangle \right) - iM \langle c^\dagger c \rangle \left\{ \langle a_c^\dagger a_c \rangle - \langle a_v^\dagger a_v \rangle \right\} \\ -iM \left\{ \langle a_c^\dagger a_c c^\dagger c \rangle^c - \langle a_v^\dagger a_v c^\dagger c \rangle^c \right\}. \qquad (4.2)$$

Compared with Eq. (2.13), a significant difference is introduced in to the dynamics of the photon-assisted polarization. The spontaneous emission source term is driven now via two expectation values: $\langle a_c^\dagger a_c \rangle$ and $\langle a_c^\dagger a_v^\dagger a_v a_c \rangle$. The second term vanishes in the OEA, since the electron cannot be in both states at the same time. Due to the presence of the WL, the OEA loses the validity and quantities with four electronic operators have to be taken into account, even without including a wetting layer - quantum dot carrier interaction in the Hamiltonian explicitly.

To close the set of differential equations, one needs to factorize the expectation values at a given level, e.g. on a Hartree-Fock level: $\langle a_i^\dagger a_j^\dagger a_k a_l \rangle \approx \langle a_i^\dagger a_l \rangle \langle a_j^\dagger a_k \rangle - \langle a_i^\dagger a_k \rangle \langle a_j^\dagger a_l \rangle$. This factorization rule is derived via the approximation, that the electron-electron correlations can be expressed as a general canonical statistical operator (GCSO), including only single particle contributions [Fri96; FS90]. That is: the single particle dynamics is described in a mean field, induced by the other particles. The approximation is done in the choice of the GCSO,[FS90] in which the observables of interest are included, e.g. ρ_{HF}:

$$\langle a_i^\dagger a_j^\dagger a_k a_l \rangle = \mathrm{tr}(a_i^\dagger a_j^\dagger a_k a_l \rho) \approx \mathrm{tr}(a_i^\dagger a_j^\dagger a_k a_l \rho_{HF}). \tag{4.3}$$

After introducing a unitary matrix $\bar{\bar{\phi}}$ to conveniently express the single particle contribution GCSO with new operators [see App. 6.4 for details]:

$$\rho_{HF} = \frac{1}{Z} e^{-\sum_n \lambda_{nn}^D d_n^\dagger d_n}, \qquad d_n^{(\dagger)} = \sum_i \phi_{in}^{(*)} a_i^{(\dagger)}, \tag{4.4}$$

inheriting the fermionic character. The partition function is calculated with the complete set of eigenfunctions of the new operators and reads:

$$Z = (1 + e^{-\lambda_{11}^D})(1 + e^{-\lambda_{22}^D}) \cdots (1 + e^{-\lambda_{NN}^D}) = \Pi_k (1 + e^{-\lambda_{kk}^D}). \tag{4.5}$$

Now, the GCSO is explicitly given and one can calculate on a Hartree-Fock level the expectation value of the electron-electron correlations:

$$\langle a_i^\dagger a_j^\dagger a_k a_l \rangle \approx \sum_{abcd} \phi_{ai} \phi_{bj} \phi_{kc}^* \phi_{ld}^* \mathrm{tr}(\rho_{HF}\, d_a^\dagger d_b^\dagger d_c d_d). \tag{4.6}$$

Only two contributions remain, if the trace is evaluated, since the states are orthogonal. The expectation value of the four electron operator quantity yields:

$$\begin{aligned}\langle a_i^\dagger a_j^\dagger a_k a_l \rangle &\approx \sum_{cd} \left[\frac{\phi_{di} e^{-\lambda_{cc}^D} \phi_{ld}^*}{(1+e^{-\lambda_{cc}^D})}\right]\left[\frac{\phi_{cj} e^{-\lambda_{dd}^D} \phi_{kc}^*}{(1+e^{-\lambda_{dd}^D})}\right] - \left[\frac{\phi_{dj} e^{-\lambda_{cc}^D} \phi_{ld}^*}{(1+e^{-\lambda_{cc}^D})}\right]\left[\frac{\phi_{ci} e^{-\lambda_{dd}^D} \phi_{kc}^*}{(1+e^{-\lambda_{dd}^D})}\right] \\ &= \langle a_i^\dagger a_l \rangle \langle a_j^\dagger a_k \rangle - \langle a_i^\dagger a_k \rangle \langle a_j^\dagger a_l \rangle. \end{aligned} \tag{4.7}$$

The calculation is straightforward, after choosing the Hartree-Fock GCSO [FS90].

4.1.2 Modified equations of motion within the SCE

Now, the equation of the photon-assisted polarization is closed on the electronic single-particle level:

$$\partial_t \langle a_v^\dagger a_c c^\dagger \rangle = -iM \langle a_c^\dagger a_c \rangle \left(1 - \langle a_v^\dagger a_v \rangle\right) - iM \langle c^\dagger c \rangle \left[\langle a_c^\dagger a_c \rangle - \langle a_v^\dagger a_v \rangle\right]$$
$$-iM \left[\langle a_c^\dagger a_c c^\dagger c \rangle^c - \langle a_v^\dagger a_v c^\dagger c \rangle^c\right]. \tag{4.8}$$

The source term of spontaneous emission is changed from being proportional to the excited state density only $\langle a_c^\dagger a_c \rangle$, to the product of ground and excited state $\langle a_c^\dagger a_c \rangle \left(1 - \langle a_v^\dagger a_v \rangle\right)$ as a result of the interaction between WL carriers and the carriers inside the QD [BGWJ06; GWLJ07]. To calculate the $g^{(2)}(t, 0)$-function, higher-order photon-assisted polarizations must be computed, considering photon-density assisted ground and excited state densities. They are not changed in comparison to the derivation within the OEA, cf. Sec. 2.3 and Eq. (2.15). The higher-order photon-assisted polarization reads within the SCE:

$$\partial_t \langle a_v^\dagger a_c c^\dagger c^\dagger c \rangle^c = \tag{4.9}$$
$$= -2iM \left[\langle a_c^\dagger a_c c^\dagger c \rangle^c (1 - \langle a_v^\dagger a_v \rangle) - \langle a_v^\dagger a_v c^\dagger c \rangle^c \langle a_c^\dagger a_c \rangle\right] + 4 \langle a_v^\dagger a_c c^\dagger \rangle^c \text{Im}\left[M \langle a_v^\dagger a_c c^\dagger \rangle^c\right]$$
$$- iM \langle c^\dagger c^\dagger c \, c \rangle^c \left[\langle a_c^\dagger a_c \rangle - \langle a_v^\dagger a_v \rangle\right] - 2i\tilde{M} \langle c^\dagger c \rangle \left[\langle a_c^\dagger a_c c^\dagger c \rangle^c - \langle a_v^\dagger a_v c^\dagger c \rangle^c\right].$$

Again, the Pauli-blocking term for the spontaneous emission is changed and now photon-assisted. The higher the order in the electron-photon coupling element, the more correction terms originate from the cluster expansion method. Those corrections lead to interesting terms in Eq. (4.9), such as the typical correlation expansion correction term $\propto |\langle a_v^\dagger a_c c^\dagger \rangle^c|^2$. The Hartree-Fock factorization in the higher-order photon-assisted polarization includes additionally a photon density assisted Pauli-blocking, cf. first line in Eq. (4.9)). But in the limit of few photons, this quantity is small in comparison to other contributions and is often neglected [GWLJ07]. Together with the equations in Sec. 2.3, a closed set of differential equations is derived and can be solved for different initial conditions of the electron and the photon system.

In Fig. 4.2(left), the Hartree-Fock contributions of the carrier-carrier interaction in the second order of the electron-light coupling element is compared with the calculation for systems, in which the OEA is valid. The excited state is assumed to be initially 1 and no photons are in the cavity. In the SCE within the OEA (dashed, red line), the vacuum Rabi oscillations lead to a negative probability for the electron density in the conduction band state, cf. Sec. 2.3. Including the enhanced Pauli-blocking due to the implicit WL interaction, the negativities are getting smaller, remain 10% large (green, solid line). Within the first oscillation (up to 250 fs), the deviation between the OEA and the QD-WL case is small and negligible. For longer times, a smaller Rabi frequency is observable for the QD-WL case, but the solution is not valid for longer times [RCSK09a]. The same initial values for the electron system is chosen in Fig. 4.2(right), but with three Fock photons in the cavity $p_3 = 1$, i.e. $\langle c^\dagger c^\dagger c c \rangle(0) = 6$ and $\langle c^\dagger c \rangle = 3$. In addition to the vacuum Rabi oscillation, the electronic system is now driven by the

4 Quantum dot - wetting layer cavity quantum electrodynamics

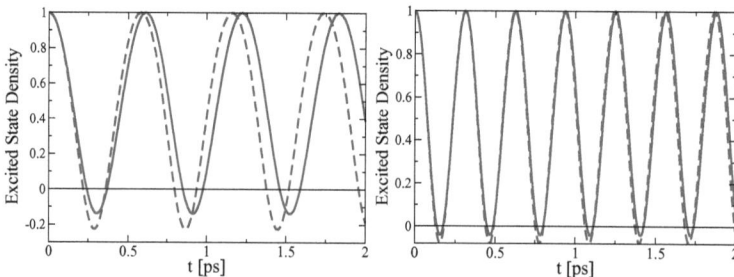

Figure 4.2: Dynamics of the excited state density. Left: For the case of a initially occupied conduction band, $\langle a_c^\dagger a_c \rangle(0) = 1$ and zero photons in the cavity, the Hartree-Fock factorization (solid, green line) results in decreased negativity in comparison to the calculation, in which carrier-carrier interaction is neglected (dashed, red line). Right: With the same electronic initial conditions but with $\langle c^\dagger c \rangle = 6$, to show, that for increasing photon number inside the cavity, first, the negativity is reduced and the difference between Hartree-Fock case and one-electron assumption becomes small.

cavity photons. Therefore, the Rabi frequency is higher. The dynamics of the OEA (dashed, red line) and QD-WL case (green, solid line) do not deviate much (percentage range). The Rabi frequency is the same and the negativities, still present, are decreased due to the higher number of photons, corresponding to a weaker correlation between electron and photon system, cf. Chap. 2. Note, the influence of the Hartree-Fock modified spontaneous emission processes is smaller and vanishes for a high number of photons.

Since negativities still occur, even within the Hartree-Fock factorization, causing a decreased oscillation amplitude, it is necessary to go beyond second order and investigate within the SCE approach the dynamics up to the fourth order. At this level, it is also possible to compute the $g^{(2)}(t, 0)$-function. The intensity-intensity correlation $\langle c^\dagger c^\dagger cc \rangle$ is driven by the photon-density assisted polarization in Eq. (4.9) and is the crucial quantity to determine the quantum light statistics of the emitted light. In Fig. 4.3, the Hartree-Fock contributions of the carrier-carrier interaction is compared to the calculations for systems, in which the OEA is valid, now up to the 4th-order in the electron-photon interaction. The initial conditions for the electron system is not changed and the excited state assumed to be initially populated and for (left panel), the cavity is empty $p_0 = 1$. The dynamics of the OEA (dashed, red line) and the QD-WL case (solid, green line) differ strongly. Negativities do not occur, but the oscillation amplitude is much too small compared to the known result of the JCM solution, cf. Sec. 2.2. The enhanced Pauli-blocking decreases the amplitude further as well as the Rabi frequency. In Fig. 4.3(right), the dynamics are investigated with three photons in the cavity. Still, the OEA and the QD-WL do not differ much. With a higher number of photons, the solution converges to the JCM solution and the carrier-carrier interaction do not lead to a different dynamics.

The calculations show, the break down of the cluster expansion approach is not compensated by the inclusion of many particle contributions in the Hartree-Fock regime. The JCM solution is reproduced only for very high numbers of photons or emitters. For few photons and few emitters, the cluster ex-

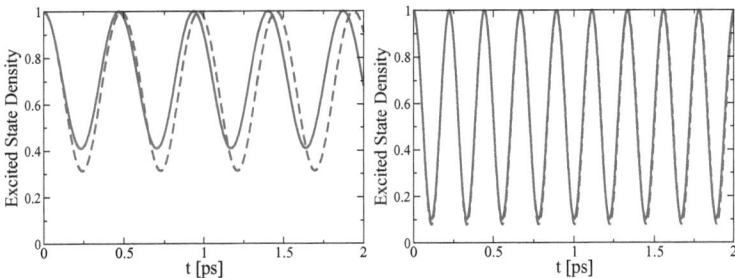

Figure 4.3: Dynamics of the excited state with higher contributions. Left: For the case of a initially occupied conduction band, $\langle a_c^\dagger a_c \rangle(0) = 1$ and zero photons in the cavity, the Hartree-Fock factorization (solid, green line) results in a decreased oscillation amplitude, it is now positive and does not oscillate between 0 and 1, like expected for the OEA case (dashed, red line). The HF does not better this result, but worsened it, by decreasing the amplitude further (solid, green line). Right: With the same electronic initial conditions but with $p_3 = 1$. The deviation from the JCM is decreased.

pansion breaks down. There is the urgent need, to introduce a theoretical framework for the Hartree-Fock factorization, which incorporates the JCM as well as the possibility to factorize the electronic field within a mean field theory.

4.2 Photon probability cluster expansion approach (PPCE)

In this section, the standard cluster expansion (SCE) is generalized to the limit of few photons in systems with few quantum confined electronic levels such as semiconductor QD-WL systems, cf. Fig. 4.1(a). So far, the SCE of $\langle a_c^\dagger a_v^\dagger a_v a_c \rangle$ provides access only to large photon number dynamics or in weak coupling-short time limit ($\hbar M \cdot t \ll 1$), where M is the electron-photon coupling constant, cf. Eq. (4.1). In strong correlated dynamics, the SCE becomes problematic, since the dynamics of relevant observables are not connected directly to the photon number state. Examples include quantities such as $\langle a_c^\dagger a_c c^\dagger c \rangle$, which include contribution from photon states with one and more photons, similarly the quantity $\langle a_c^\dagger a_c c^\dagger c^\dagger c c \rangle$, which includes contributions from two and more photons. However, it is clear that this kind of expansion is not suitable, if correlations with a distinct number of photons are of crucial importance.

A theoretical scheme, which attacks this problem for a single photon mode, but is easy to extend to multiple photon modes, is presented in this section. This new approach allows to treat few photon dynamics in many particle systems such as QD-WL systems with few electronic levels, that contribute to the emission. This section is organized as follows: First, the photon probability expansion is derived, cf. Sec. 4.2.1, expressing relevant observables in terms of photon probabilities. In this probability picture, the JCM solution can be analytically re-derived and the inclusion of photon-statistics is discussed, cf. Sec. 4.2.2. Then, a modified Hartree-Fock factorization rule is introduced in Sec. 4.2.3 and the modified equations of motion are investigated, cf. Sec. 4.2.4.

4.2.1 Photon probabilities expansion

Important quantities to characterize a quantum optical field are the normalized intensity-intensity correlation function $g^{(2)}(t,\tau=0) = \frac{\langle c^\dagger c^\dagger cc\rangle}{\langle c^\dagger c\rangle^2}$ and the photon intensity $g^{(1)}(\tau=0) = \langle c^\dagger c\rangle$ [MW95]. Both quantities can be measured and they are directly connected to the photon number $\langle c^\dagger c\rangle$ and intensity-intensity expectation value $\langle c^\dagger c^\dagger cc\rangle$. The idea to circumvent the problems characteristic for the SCE (cf. Sec. 4.1) is to formulate all quantities in terms of n photon probabilities $p_n = \langle |n\rangle\langle n|\rangle$, where $|n\rangle$ denotes the Fock-state of n photons in the system. This idea is already succesfully applied in JCM, but needs to be generalized to the many electron case. The PPCE introduced now combines the advantages of the exact JCM-solution and the standard cluster expansion. Rewriting the observables by introducing a complete set of eigenfunctions, thus inserting an identity of the photon subspace:

$$\mathbb{1} = \sum_n |n\rangle\langle n|. \tag{4.10}$$

The photon density reads:

$$\langle c^\dagger c\rangle = \langle c^\dagger c \mathbb{1}\rangle = \sum_n \langle c^\dagger c |n\rangle\langle n|\rangle = \sum_n n\,\langle |n\rangle\langle n|\rangle. = \sum_n n\, p_n. \tag{4.11}$$

To calculate the photon density, one must solve an equation of motion for the photon probability. The whole description of the combined electron and photon dynamics are transfered onto the photon probabilities, which is comparable to the situation in the JCM. In the JCM, the equations of motion are solved exactly for a single-photon subspace, in which a photon is emitted or absorbed from the state $|n\rangle$, cf. Sec. 2.2. So, instead of calculating the whole photon hierarchy:

$$\langle c^\dagger c\rangle \rightarrow \langle c^\dagger c^\dagger cc\rangle \rightarrow \langle c^\dagger c^\dagger c^\dagger ccc\rangle \rightarrow \cdots \tag{4.12}$$

One calculates different photon probability subspaces. In this way, the calculation is organized and can be truncated safely, since it is easily to show, that for fixed initial conditions (without environment coupling), certain photon probabilities are not driven. Therefore, it exists an N from which on the photon probability p_n is zero for all times, if no external pump mechanism is included.

The equation of motion for the p_n is derived via the Hamiltonian in Eq. (4.1) with the Heisenberg equation of motion:

$$\partial_t p_n = -2\sqrt{n}\,\text{Im}\left[M\,\langle |n\rangle\langle n-1|a_v^\dagger a_c\rangle\right] + 2\sqrt{n+1}\,\text{Im}\left[M\langle |n+1\rangle\langle n|a_v^\dagger a_c\rangle\right]. \tag{4.13}$$

To calculate the photon probability, the photon-assisted polarization enters the system dynamics. Here, the single-electron case is considered, assuming quantities like $\langle a_i^\dagger a_j^\dagger a_k a_l\rangle$ to be zero for all time and combination of $i,j,k,l = \{v,c\}$.

$$\partial_t \langle a_v^\dagger a_c |n+1\rangle\langle n|\rangle = -i\,M\,\sqrt{n+1}\left(\langle a_c^\dagger a_c |n\rangle\langle n|\rangle - \langle a_v^\dagger a_v |n+1\rangle\langle n+1|\rangle\right). \tag{4.14}$$

This form of the photon-assisted polarization includes the induced absorption and emission processes, as well as the spontaneous emission. In Eq. (4.14) a new form of the phase-filling factor is derived. The spontaneous emission and induced emission is included in the quantity $\langle a_c^\dagger a_c |n\rangle\langle n|\rangle$, whereas the induced absorption is included in $\langle a_v^\dagger a_v |n+1\rangle\langle n+1|\rangle$. To see that, one can transform Eq. (4.14) back into the photon operator picture. Details are found in App. 6.5. The photon-assisted valence and conduction band state densities read:

$$\partial_t \langle a_v^\dagger a_v |n\rangle\langle n|\rangle = -2\sqrt{n}\, \mathrm{Im}\!\left(M\, \langle a_v^\dagger a_c |n\rangle\langle n-1|\rangle\right) \quad (4.15)$$

$$\partial_t \langle a_c^\dagger a_c |n\rangle\langle n|\rangle = 2\sqrt{n+1}\, \mathrm{Im}\!\left(M\, \langle a_v^\dagger a_c |n+1\rangle\langle n|\rangle\right). \quad (4.16)$$

In this expansion, the photon-assisted valence and conduction band density dynamics differ strongly. This can be attributed to the rotating-wave approximation, already done in the Hamiltonian. This results from the density matrix description, and that the valence state is driven by a lower order of the photon-assisted polarization. Instead of the electron-picture, one can formulate the same equations in the electron-hole picture, taking as photon-assisted electron density $f_n^e = \langle a_c^\dagger a_c |n\rangle\langle n|\rangle$ and as photon-assisted hole density $f_n^h = \langle |n\rangle\langle n|\rangle - \langle a_v^\dagger a_v |n\rangle\langle n|\rangle$. In this picture, the photon-probability reamains the same, but the photon-assisted polarization reads now:

$$\partial_t \langle a_v^\dagger a_c |n+1\rangle\langle n|\rangle = +i\, M\, \sqrt{n+1}\,(p_{n+1} - f_{n+1}^h) - i\, M\, \sqrt{n+1}\, f_n^e. \quad (4.17)$$

Quantum optical absorption and emission processes are related to $f_n^h = p_n - \langle |n\rangle\langle n|a_v^\dagger a_v\rangle$ and $f_n^e = \langle |n\rangle\langle n|a_c^\dagger a_c\rangle$ for holes and electron densities assisted by n photons. Finally, the electron and hole densities $f_n^{e/h}$ are given by:

$$\partial_t f_n^{e/h} = 2\sqrt{n+1}\, \mathrm{Im}\!\left(M\, \langle a_v^\dagger a_c |n+1\rangle\langle n|\rangle\right). \quad (4.18)$$

In the electron-hole picture, the dynamics of the photon-assisted electron and hole dynamics obey the same equations of motion.

To close the hierarchy of equations of motion, $\langle a_i^\dagger a_j^\dagger a_k a_l |n\rangle\langle n|\rangle = 0$ is used and is strictly valid only for the single electron case, since two electrons are annihilated. Eq. (4.13) and Eqs. (4.17)-(4.18)) reproduce the JCM, numerically and analytically. This is an important benchmark of the new PPCE, if the model is limited to one electron in the quantum dot, see App. 6.6.

4.2.2 PPCE and photon-statistics

Via the choice of initial conditions for the cavity field, the JCM calculates the dynamics for different photon-statistics, such as thermal light and coherent light, cf. Sec. 2.2. In the PPCE, the expectation values are expressed with the statistical properties, which is one of the advantages of this approach. For example, the photon density is calculated with:

$$\langle c^\dagger c \rangle = \sum_n n \langle |n\rangle\langle n| \rangle = \sum_n n\, p_n. \tag{4.19}$$

The whole information of the quantum correlations are incorporated in p_n. Different light fields are characterized via the deviations around the mean photon number. Even with a given mean photon number, e.g., the standard deviation can be different. Important examples are thermal and coherent distributions, cf. Sec. 3.2.3.

The mean photon number of a thermal light field is $\bar{n} = \langle c^\dagger c \rangle$ with a probability distribution:

$$p_n = \frac{\bar{n}^n}{(\bar{n}+1)^{n+1}}. \tag{4.20}$$

To check this approach and the right choice of the statistical operator, the mean photon number is calculated:

$$\langle c^\dagger c \rangle = \sum_n n\, p_n = \sum_{n=0}^{\infty} n \frac{\bar{n}^n}{(\bar{n}+1)^{n+1}} = \left(\frac{\bar{n}}{\bar{n}+1}\right)\left(\frac{\bar{n}}{\bar{n}+1}\right)^{-1} \frac{1}{\bar{n}+1} \sum_n n \left(\frac{\bar{n}}{\bar{n}+1}\right)^n.$$

It is convenient to abbreviate $q = \frac{\bar{n}}{\bar{n}+1} < 1$:

$$\langle c^\dagger c \rangle = \frac{q}{\bar{n}+1} \sum_n n\, q^{n-1} = \frac{q}{\bar{n}+1} \partial_q \left[\sum_n q^n\right] = \frac{q}{\bar{n}+1} \partial_q \left[\frac{1}{1-q}\right] \tag{4.21}$$

$$= \frac{q}{\bar{n}+1}\left[\frac{1}{1-q}\right]^2 = \frac{\bar{n}}{(\bar{n}+1)^2}\left[\frac{1}{1-\frac{\bar{n}}{\bar{n}+1}}\right]^2 = \frac{\bar{n}}{(\bar{n}+1)^2}[\bar{n}+1]^2 = \bar{n}.$$

The thermal light cavity dynamics are initialized with the given p_n, also reproducing the JCM. Another example is the coherent state, with a probability distribution, taken into account the Poissonian statistics of the coherent light field:

$$p_n = \frac{\bar{n}^n}{n!} e^{-\bar{n}}, \tag{4.22}$$

here $\bar{n} = |\alpha|^2$ corresponds to the mean photon number of the Glauber states. Again, the expectation value is calculated:

$$\langle c^\dagger c \rangle = e^{-\bar{n}} \sum_{n=1}^{\infty} n \frac{\bar{n}^n}{n!} = e^{-\bar{n}} \sum_{n=1}^{\infty} \frac{\bar{n}^n}{(n-1)!} = \bar{n}\, e^{-\bar{n}} \sum_{n=1}^{\infty} \frac{\bar{n}^{n-1}}{(n-1)!}, \tag{4.23}$$

now the summation index changed into $m = n - 1$:

$$\langle c^\dagger c \rangle = \frac{\bar{n}}{e^{\bar{n}}} \sum_{m=0}^{\infty} \frac{\bar{n}^m}{m!} = \frac{\bar{n}}{e^{\bar{n}}} e^{\bar{n}} = \bar{n}. \tag{4.24}$$

4 Quantum dot - wetting layer cavity quantum electrodynamics

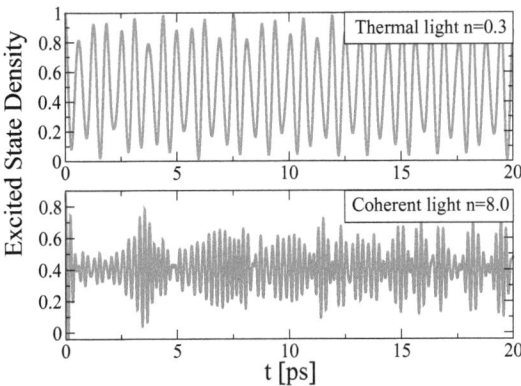

Figure 4.4: Photon-statistics within the PPCE: thermal light and coherent light dynamics for a two-level system, initially prepared in the excited state. The solution of the JCM is reproduced. Superposition of Rabi oscillation lead to a complex oscillation pattern for thermal light, and to the phenomenon of collapse and revival for coherent light.

In Fig. 4.4, numerical evaluations of thermally and coherently prepared cavity dynamics are plotted. The JCM solutions from Sec. 2.2 is reproduced within the PPCE approach. Depending on the mean photon number, higher order of the n-photon probabilities need to be taken into account.

However, the PPCE approach is not limited to the restrictions of the JCM, since it is now possible to incorporate, e.g., enhanced Pauli-blocking due to carriers scatter in and out of the quantum dot and violating the OEA.

4.2.3 PPCE and Hartree Fock factorization

In this section, a factorization rule for the photon-assisted electron-electron correlation is derived. After expanding the photon system in terms of photon-probabilities $p_n = \langle |n\rangle\langle n| \rangle$, the equation of motion of the photon-assisted polarization beyond the OEA reads:

$$\partial_t \langle a_v^\dagger a_c |n+1\rangle\langle n| \rangle \quad (4.25)$$
$$= -i M \sqrt{n+1}\, \langle a_c^\dagger a_c |n\rangle\langle n| \rangle + i M \sqrt{n+1}\, \langle a_v^\dagger a_v a_c |n\rangle\langle n| \rangle$$
$$+ i M \sqrt{n+1}\, \langle a_v^\dagger a_v |n+1\rangle\langle n+1| \rangle - i M \sqrt{n+1}\, \langle a_v^\dagger a_c^\dagger a_c a_v |n+1\rangle\langle n+1| \rangle.$$

In Eq. (4.25), electron-electron correlations enter via e.g. $\langle a_v^\dagger a_c^\dagger a_c a_v |n\rangle\langle n| \rangle$. To close the set of differential equations, one needs to factorize electron correlations expectation values. For example, in the lowest order, one describes a two-particle quantity via the dynamics of electron single particle quantities. This is a mean field theory, in which a single carrier is influenced by a mean field. This mean field is the first approximation to include carrier-carrier interaction, here on a Hartree-Fock level. The usual Hartree-Fock approximation is not applicable here, since the photon contributions are neither part of the one electron operator pair nor of the other. The Hartree-Fock factorization needs to be modified according to the photon dynamics, and photon expansion technique. Starting with the expectation value, which needs to be factorized, one derives, given a state with $n_1 \cdots n_N$ electrons and m photons:

$$\begin{aligned}\langle |n\rangle\langle n|a_i^\dagger a_j^\dagger a_k a_l\rangle &= \sum_{m,\{n_i\}} \langle n_1...n_N,m|\,|n\rangle\langle n|a_i^\dagger a_j^\dagger a_k a_l \rho |n_1...n_N,m\rangle \\ &\approx \sum_{\{n_i\}} \langle n_1...n_N|a_i^\dagger a_j^\dagger a_k a_l \langle n|\rho|n\rangle |n_1...n_N\rangle.\end{aligned} \quad (4.26)$$

In the last line, the approximation is applied, that the photon state $|n\rangle$ does not include any electronic contributions, for all times. In other words, the electron and photon system on each photon-probability p_n is in a quasi-equilibrium. Note, this is not equivalent to the Born factorization. The statistical operator does not factorize. The GCSO includes every contribution, in dependence on the choice of the observable set. Here, the GCSO is projected onto the different photon-probabilities, resulting in the PPCE - GCSO:

$$\rho_{pn} = \frac{1}{p_n} \langle n|\rho|n\rangle. \quad (4.27)$$

With p_n in the denominator, the PPCE-GCSO is trace conserving. To illustrate this, an example is given. A simple case is considered, in which the electron ρ_{el} and photon system $\rho_{pt} = \sum_m p_m |m\rangle\langle m|$ statistical operator are factorizing completely:

$$\begin{aligned}\rho_{pn} &= \frac{1}{p_n}\langle n|\rho|n\rangle = \frac{1}{p_n}\langle n|\rho_{pt}\otimes\rho_{el}|n\rangle = \frac{1}{p_n}\langle n|\sum_m p_m |m\rangle\langle m|\otimes \rho_{el}|n\rangle \\ &= \frac{1}{p_n}\sum_m p_m \langle n|m\rangle\langle m|\rho_{el}|n\rangle = \frac{p_n}{p_n}\langle n|n\rangle\rho_{el} = \rho_{el}.\end{aligned} \quad (4.28)$$

The Hartree-Fock GCSO now has a different form, assuming the electron-electron correlations are describable on a single-particle basis, and this, for every photon-probability, the GCSO reads:

$$\rho \approx \rho_{HF} := \frac{e^{-\sum_{ij}\lambda_{ij}a_i^\dagger a_j}}{Z} \to \rho_{pn} = \frac{\langle n|\rho|n\rangle}{p_n} \approx \frac{1}{Z}e^{-\sum_{ij}\lambda_{ij}^n a_i^\dagger a_j} := \sigma_{el}^n. \quad (4.29)$$

With this statistical operator (σ_{el}^n), the usual Hartree-Fock factorization, cf. App. 6.4, is calculated straightforward. Therefore, one introduces a p_n/p_n to realize this normalized GCSO in Eq. (4.26) to yield:

$$\langle |n\rangle\langle n| a_i^\dagger a_j^\dagger a_k a_l \rangle \tag{4.30}$$

$$\approx p_n \sum_{\{n_i\}} \langle n_1,...n_N | a_i^\dagger a_j^\dagger a_k a_l \sigma_{el}^n | n_1...n_N \rangle$$

$$= p_n \sum_{cd} \left[\frac{\phi_{di} e^{-\lambda_{cc}^{n,D}} \phi_{ld}^*}{(1 + e^{-\lambda_{cc}^{n,D}})}\right]\left[\frac{\phi_{cj} e^{-\lambda_{dd}^{n,D}} \phi_{kc}^*}{(1 + e^{-\lambda_{dd}^{n,D}})}\right] - \left[\frac{\phi_{dj} e^{-\lambda_{cc}^{n,D}} \phi_{ld}^*}{(1 + e^{-\lambda_{cc}^{n,D}})}\right]\left[\frac{\phi_{ci} e^{-\lambda_{dd}^{n,D}} \phi_{kc}^*}{(1 + e^{-\lambda_{dd}^{n,D}})}\right],$$

with $\lambda_{ii}^{n,D}$ as the matrix element of the diagonalized matrix for the n-th photon-probability. One calculates now with the same equilibrium assumption and introduced GCSO the single electron quantity, to identify the contributions:

$$\langle |n\rangle\langle n| a_i^\dagger a_l \rangle \approx \sum_{\{n_i\}} p_n \langle n_1...n_N | a_i^\dagger a_l \sigma_{el}^n | n_1...n_N \rangle = \sum_d \left[p_n \frac{\phi_{di} e^{-\lambda_{cc}^{n,D}} \phi_{ld}^*}{(1 + e^{-\lambda_{cc}^{n,D}})} \right]. \tag{4.31}$$

Now, in Eq.(4.30), one can identify the modified Hartree-Fock rule, by inserting again p_n/p_n to take into account the p_n contribution to the single particle quantity in Eq. (4.31):

$$\langle |n\rangle\langle n| a_i^\dagger a_j^\dagger a_k a_l \rangle \tag{4.32}$$

$$\approx \frac{1}{p_n} \sum_{cd} \left[p_n \frac{\phi_{di} e^{-\lambda_{cc}^{n,D}} \phi_{ld}^*}{(1 + e^{-\lambda_{cc}^{n,D}})}\right]\left[p_n \frac{\phi_{cj} e^{-\lambda_{dd}^{n,D}} \phi_{kc}^*}{(1 + e^{-\lambda_{dd}^{n,D}})}\right] - \left[p_n \frac{\phi_{dj} e^{-\lambda_{cc}^{n,D}} \phi_{ld}^*}{(1 + e^{-\lambda_{cc}^{n,D}})}\right]\left[p_n \frac{\phi_{ci} e^{-\lambda_{dd}^{n,D}} \phi_{kc}^*}{(1 + e^{-\lambda_{dd}^{n,D}})}\right]$$

$$= \frac{1}{p_n} \left\{ \langle |n\rangle\langle n| a_i^\dagger a_l \rangle \langle |n\rangle\langle n| a_j^\dagger a_k \rangle - \langle |n\rangle\langle n| a_i^\dagger a_k \rangle \langle |n\rangle\langle n| a_j^\dagger a_l \rangle \right\},$$

which defines the factorization rule, taken into account semiconductor many-particle contributions, resulting from the independent hole and electron dynamics with valence and conduction band state filling. The factorization rule originates from the assumption that the photon and electron dynamics on every photon number is approximately described as in equilibrium and thus with the statistical operator shown in Eq.(4.29). The GCSO does not imply, that the total photon and electron dynamics are in equilibrium, which would lead to a factorized statistical operator $\rho = \rho_{pt} \otimes \rho_{el}$ and result in a Born factorization rule:

$$\langle |n\rangle\langle n| a_i^\dagger a_j^\dagger a_k a_l \rangle \approx \langle |n\rangle\langle n| \rangle \langle a_i^\dagger a_j^\dagger a_k a_l \rangle. \tag{4.33}$$

This is only valid in case of a weak coupling, or a weak correlation between the photon and electron dynamics, e.g. if the photons are treated as a bath [VWW01]. This assumption is not valid in the case of a single QD coupled to single cavity mode in the strong coupling limit, interesting for the deterministic single photon emission on demand [MKB+00].

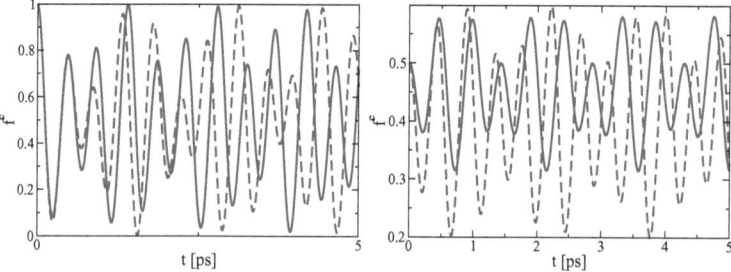

Figure 4.5: Dynamics of f^e driven by cavity photons. Left: For the case of a initially occupied conduction band state, $f^e = f^h = 1$ and $p_0 = 0.4, p_1 = 0.6$ photons in the cavity, the Hartree-Fock factorization (solid, green line) results after 1 ps in a different dynamics, compared to the OEA (dashed, red line). Right: With $f^e = f^h = 0.5$, the enhanced Pauli-blocking is clearly visible. The oscillation amplitude is decreased, if the Hartree-Fock factorization is included (green, solid line).

4.2.4 PPCE and modified equations of motion

The equation of motion, describing the combined and strongly coupled electron-photon dynamics, are modified due to the many-particle contributions in contrast to the equations, given within the mathematical induction approach in Chap. 3. Since in shallow QD electrons and holes in the wetting layer interact with the carriers inside the QD, the polarization dynamics of the QD is modified. New contributions occur in comparison to the photon-assisted polarization in the one-electron assumption, Eq. (4.32), namely in the braces. They lead to an enhanced Pauli-blocking. They decrease the impact of the photon-assisted electron densities. In the electron-hole picture, [for the electron picture, cf. App. 6.7], the equation reads:

$$\partial_t \langle a_v^\dagger a_c | n+1 \rangle \langle n | \rangle = -i M \sqrt{n+1} \left[\frac{f_n^h f_n^e}{p_n} - \frac{(p_{n+1} - f_{n+1}^h)(p_{n+1} - f_{n+1}^e)}{p_{n+1}} \right]. \tag{4.34}$$

The electron densities and the photon densities stay unchanged. The set of equations of motion can be evaluated numerically for given initial conditions.

To investigate the impact of the modified spontaneous emission and induced absorption and emission processes, the dynamics of the electron densities are depicted in Fig. 4.5 in the OEA (dashed, red line) and including the WL Pauli-blocking (solid, green line) the photon probabilities are $p_0 = 0.4, p_1 = 0.6$. In Fig. 4.5(left), the electron densities are set initially $f^e = f^h = 1$. Since higher order photon-assisted polarization are driven with the photon-probabilities, the Pauli-blocking modifies the dynamics and decreases the Rabi frequency. Higher order photon-probabilities lead to a collapse and revival similiar dynamics, which is enhanced due to the Pauli-blocking. In Fig. 4.5(right), the electron densities are set initially $f^e = f^h = 0.5$ and the picture does not change but the amplitude. The electron density does not oscillate between 0.2 and the initial value 0.7, like in the case, in which the OEA (dashed, red line) is valid. The amplitude is decreased due to the interplay between hole and electron density with the independent carrier interaction of the WL. As an additional important result, the Rabi frequency is also different. The Pauli blocking decreases the Rabi frequency (solid, green line) although the amplitude decreased. This is clearly a situation, which goes beyond the JCM, cf. Sec. 2.2. In the OEA, a decreased amplitude lead to a higher Rabi frequency, e.g. in case of a detuning, cf. Sec. 3.2.2. Here, the amplitude is decreased as well as the Rabi frequency. After 2 ps, the maximum of the dynamics are clearly shifted. If more than one photon probabilities are inequal to zero in the beginning, the Pauli-blocking factors modify the dynamics due to the denominator in Eq. (4.17).

In contrast to the JCM, the strength of the vacuum Rabi flopping is reduced in dependence on how strong the electron and hole density deviates from the OEA $f^h = f^e$. This implies that the amplitude of the Rabi flops might be used as a measure for the number of electrons in the actual quantum dot. To connect this result to an easily accessible quantity, the $g^{(2)}$-function is calculated in the next section and the dynamics of the SCE and the PPCE are compared.

4.3 Photon dynamics in the PPCE and SCE approach

The best benchmark for the new PPCE scheme is the case of vacuum Rabi flops, where quantum fluctuations dominate the dynamics. The PPCE provides physical reasonable results as opposed to the SCE method: it guarantees positivity of the electronic density, cf. Sec. 4.2.4. In particular, the exact solution of the JCM is recovered within the OEA for arbitrary initial conditions, cf. Sec. 4.2. In this section, the differences between the SCE and PPCE are discussed within and beyond the validity of the OEA, focusing on the photon dynamics. The photon dynamics determine the easily accessible intensity correlation $g^{(2)}(t,0)$-function and the photon density. Both depend strongly on the electron dynamics discussed in the previously section. In Sec. 4.3.1, the differences within the OEA are discussed, in which the PPCE-solution is identical with the JCM solution and is used as a benchmark to determine to what extent the SCE is a good approximation in a QD-WL system. Here, the investigation focuses on the $g^{(2)}(t,0)$-function. In Sec. 4.3.2, the results of the PPCE are applied to extract the actual number of electrons in the QD from the Rabi oscillation amplitude of the $g^{(2)}(t,0)$-function.

4.3.1 Intensity-intensity correlations within the Hartree-Fock approximation

The photon density, identical to the $g^{(1)}(t, \tau = 0)$, and the $g^{(2)}(t,0)$-function, are very interesting quantities for technological applications, and are well-known in atomic quantum optics, in which the OEA is valid and results are known. To calculate the second-order correlation $g^{(2)}(t,0)$-function within the SCE, one has to determine the correction quantity $\langle c^\dagger c^\dagger cc \rangle^c$, which is defined as the difference between the full correlation $\langle c^\dagger c^\dagger cc \rangle$ and the sum of all possible factorizations, cf. Sec. 2.3:

$$\langle c^\dagger c^\dagger cc \rangle^c = \langle c^\dagger c^\dagger cc \rangle - 2(\langle c^\dagger c^\dagger c \rangle^c \langle c \rangle^c + \langle c^\dagger cc \rangle^c \langle c^\dagger \rangle^c + \langle c^\dagger c \rangle^c \langle c^\dagger c \rangle^c + \langle c^\dagger c \rangle^c \langle c^\dagger \rangle^c \langle c \rangle^c)$$
$$- \langle c^\dagger c^\dagger \rangle^c \langle c \rangle^c \langle c \rangle^c - \langle cc \rangle^c \langle c^\dagger \rangle^c \langle c^\dagger \rangle^c - \langle c^\dagger c^\dagger \rangle^c \langle cc \rangle^c - \langle c^\dagger \rangle^c \langle c^\dagger \rangle^c \langle c \rangle^c \langle c \rangle^c$$

which simplifies in case of Fock states in the cavity to

$$\langle c^\dagger c^\dagger cc \rangle^c = \langle c^\dagger c^\dagger cc \rangle - 2 \left(\langle c^\dagger c \rangle^c \right)^2. \tag{4.35}$$

Other correlations are not driven and have an initial value of zero. In this case, the photon density is equal to the correlation: $\langle c^\dagger c \rangle = \langle c^\dagger c \rangle^c$. This is not valid, if coherent states are assumed within the cavity, i.e. Glauber states, which have non-vanishing singlet expectation values: $\langle c^\dagger \rangle$ [Gla63]. The $g^{(2)}(t,0)$-function is calculated in the SCE approach with:

$$g^{(2)}(t, \tau = 0) = \frac{2\langle c^\dagger c \rangle^2 + \langle c^\dagger c^\dagger cc \rangle^c}{\langle c^\dagger c \rangle^2} = 2 + \frac{\langle c^\dagger c^\dagger cc \rangle^c}{\langle c^\dagger c \rangle^2}. \tag{4.36}$$

Since the $g^{(2)}$-function depends on the preparation of the cavity field, the correction term depends on the initial conditions and has to be calculated with:

$$\langle c^\dagger c^\dagger cc \rangle(0) - 2\langle c^\dagger c \rangle^2(0) = \langle c^\dagger c^\dagger cc \rangle^c(0) = (g^{(2)}(0) - 2) \langle c^\dagger c \rangle^2(0). \tag{4.37}$$

Now, with a given set of initial conditions the equations of motion are evaluated and the differences between the SCE and PPCE can be discussed with the OEA: $\langle a_c^\dagger a_c \rangle = 1 - \langle a_v^\dagger a_v \rangle$, which leads to the specific relation within the SCE, already mentioned and discussed: $\langle a_c^\dagger a_c c^\dagger c \rangle^c = -\langle a_v^\dagger a_v c^\dagger c \rangle^c$.
For a high number of emitters, or a high number of photons inside the cavity, the SCE and the PPCE converge, since for a high number of photons the SCE becomes a very good approximation technique and is well established. In Fig. 4.6(left), the dynamics of the photon density and the $g^{(2)}(t,0)$-function is depicted for a high number of photons initially inside the cavity and within the OEA. In this case, the SCE (dashed, orange line) and the PPCE (solid, green line) converge. The higher the number of photons, the better the SCE approximates the system, and the closer the SCE reproduces the JCM solution, here derived within the PPCE. The photon density oscillates maximally between 40 and 41 with one QD in the cavity. The QD is initially in the excited state. However, the Rabi frequency of the SCE solution deviates from the value of the exact model. The SCE Rabi frequency is higher, approximately 1/3-of the JCM Rabi frequency, given by $\Omega = M\sqrt{N+1} = M\sqrt{41}$. This would be visibile in a calculated spectrum, but does not change the photon-statistics of the quantum light emission from the QD, which is almost 1 for both solutions. The deviation is in a percentage-range, which is small, compared to the value of 1. The mean value for the SCE is 1.0, the mean value of

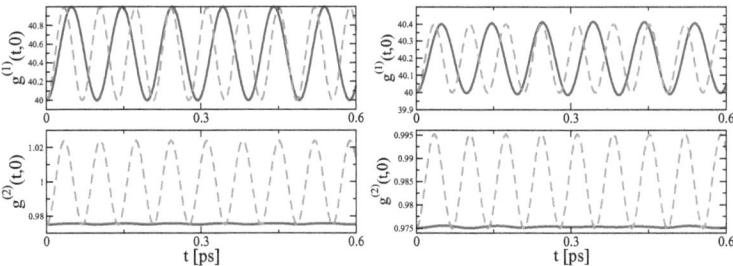

Figure 4.6: Dynamics of the photon density $g^{(1)}(t,0)$ and the intensity-intensity correlation function $g^{(2)}(t,0)$, calculated with the SCE (dashed, orange line) and the PPCE (solid, green line) in the one-electron approximation (left) and for a QD-WL system (right). The QD is initially excited and inside the cavity are $g^{(1)}(0) = 40$ photons. The solutions differ not much, but the Hartree-Fock factorization within the SCE results in a too high Rabi frequency. The enhanced Pauli-blocking leads to a decreased oscillation amplitude for the oscillation of the photon density and to a decreased photon-statistics.

the PPCE almost 0.98. For a even higher number, the mean value of the PPCE slowly converges $g^{(2)}(t,0) \to 1$, since the correlation function can be calculated for a Fock state with:

$$g^{(2)}(t,0) = 1 - \frac{1}{\langle c^\dagger c \rangle(t)}, \quad (4.38)$$

which gives $g^{(2)} = 0.975$ for $N = 40$, which is exactly the numerical value, visible in Fig. 4.6(left). From about $N = 100$, the solution become the same, within less than one percent. In Fig. 4.6(right), the initial conditions for the photon system are the same, but the electronic system is in the state $f^e = 0.8$ and $f^h = 0.4$, to emphasize the QD-WL properties for typical initial conditions, beyond the OEA. Again, the SCE (dashed, orange line) and the PPCE solutions (solid, green line) are compared. The difference is small due to the high number of photons. One photon more or less does not change the electron-photon interaction much. However, the Hartree-Fock factorization does not solve the problem of a too high Rabi frequency in the SCE solution (dashed, orange line). Interestingly, the enhanced Pauli-blocking leads to a decreased oscillation amplitude of the photon density, which here is emphasized by the choice of the initial conditions. Moreover, the photon statistics is decreased also, in particular for the SCE solution. Without the Hartree-Fock Pauli-blocking term, the photon-statistics oscillates between 1.02 and 0.98, with the Pauli-blocking term in the spontaneous and induced emission, the $g^{(2)}$-function oscillates only between 0.995 and 0.987. The SCE solution is closer to the PPCE solution now. The Pauli-blocking from the many-particle carrier-carrier interaction leads to decreased $g^{(2)}(t,0)$-function. The fundamental problem with this theoretical approach cannot be solved and becomes more obvious and problematic, if the QD is initially in the ground state.

In Fig. 4.7(left), the photon density and intensity-intensity correlation function is depicted for a QD initially in the ground state and with two photons inside the cavity. The SCE shows for the photon density (dashed, orange line) a strong deviation from the JCM solution. The oscillation amplitude is

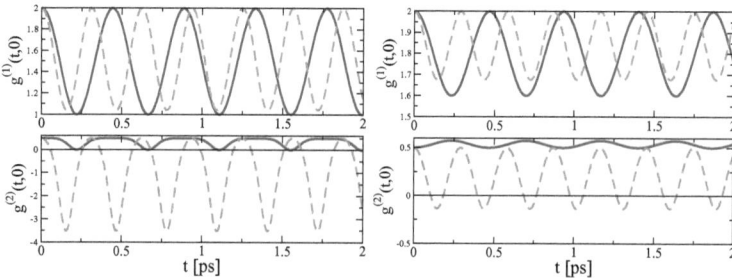

Figure 4.7: Dynamics of the photon density $g^{(1)}(t,0)$ and the intensity-intensity correlation function $g^{(2)}(t,0)$, calculated with the SCE (dashed, orange line) and the PPCE (solid, green line) in the one-electron approximation (left) and for a QD-WL system (right). The QD is initially in the ground state. The Pauli-blocking leads to a decreased oscillation amplitude for both solution, PPCE and SCE. In the SCE solution, negativities occur in the $g^{(2)}(t,0)$-function oscillation.

decreased, but now, the photon density does not reach 1 and does not create a full excitation within the system in contrast to the PPCE solution (solid, green line), which correspond to the expected behavior. The decreased amplitude leads now for the intensity-intensity correlation function to a obvious failure of the SCE, since the positively defined correlation function becomes negative and oscillates between 0.5, which is the expected maximum, and −3, which exceeds the minimum 0 of the JCM solution. Even without a benchmark, this failure leads to the conclusion, that within the OEA, the SCE is valid only for a high number of emitters, whereas the PPCE provides a theoretical framework to rely on. In Fig. 4.7(right), the QD-WL case is investigated with a QD initially in $f^e = 0.4$ and $f^h = 0.2$. Two photons are inside the cavity. The Pauli-blocking factor leads to the decrease in the oscillation amplitude, now for both, the PPCE and the SCE solution with the result, that the negativities in the $g^{(2)}(t,0)$-function become smaller for the SCE. The PPCE secures the positivity of the $g^{(2)}(t,0)$-function, as long as the limit of the Hartree-Fock approximation is valid. The comparison now favours clearly in the single-photon limit, with few emitters and few photons, the theoretical approach of the PPCE.

4.3.2 Pauli-blocking dependent Rabi oscillation amplitude

The Pauli blocking term occurring in the QD-WL system does not lead to strong deviations from the JCM solution for short times. The impact of the Pauli-blocking terms depends strongly on the number of electrons and holes in the QD, without environment coupling, e.g. electrical pumping. Since the $g^{(2)}(t,0)$-function is an easy accessible observable and the time evolution is, in principle, measurable, the amplitude can be used as a measure for the number of electrons and holes in the actual QD for the QD-WL system. In those systems, the oscillation amplitude differs for different initial conditions in the electronic system strongly.

Focusing on the specific QD-WL properties, in Fig. 4.8, the dynamics of the electron density (dashed,

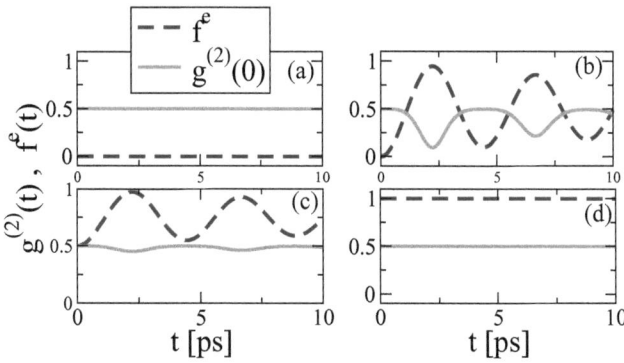

Figure 4.8: Dynamics of the $g^{(2)}(t,0)$-function (solid, orange line) and the electron density (dashed, blue line) for different initial conditions of the electronic system $N_E = 1 + f^e - f^h$ and initially two photons in the cavity. (a) $N_E = 0$: No dynamics for the $g^{(2)}(t,0)$-function, since no electrons in the QD; (b) $N_E = 1$: Cavity dynamics lead to strong Rabi oscillations in the one-electron limit $f^h = f^e$; (c) $N_E = 1.5$: the number of electron and holes are increased, the oscillation amplitude is smaller due to Pauli-blocking; and (d) $N_E = 2$: Full Pauli-blocking prevent cavity dynamics.

blue line) and the $g^{(2)}(t,0)$-function (solid, orange line) is depicted for different initial conditions in the electronic field, the number of electrons and holes in the QD with $N_E = 1 - f^h + f^e = \langle a_v^\dagger a_v \rangle + \langle a_c^\dagger a_c \rangle$. The cavity has two photons in the beginning, i.e. $p_2 = 1$ and the photon-statistics is anti-bunched:

$$g^{(2)}(t=0,0) = \frac{\sum_n n(n-1)p_n}{(\sum_n np_n)^2} = \frac{2}{(2)^2} = \frac{1}{2}. \quad (4.39)$$

Via the choice of different initial values for the total number of electrons and holes in the QD, the impact of the Pauli-blocking on the oscillation behavior can be investigated. In Fig. 4.8(a), there is no electron in the QD and no dynamics takes place. The photon-statistics is constantly $g^{(2)} = 0.5$ and the electron density $f^e = 0$. In Fig. 4.8(b), the total number is increased to $N_E = 1$ by setting $f^h = f^e = 0$, or assuming $\langle a_v^\dagger a_v \rangle = 1$. Dynamics take place, the electron density oscillates strongly with an amplitude of nearly 1, and the $g^{(2)}(t,0)$-function follows with an oscillation amplitude of about 0.3. For a higher total number of electrons and holes, $N_E = 1.5$, or $\langle a_c^\dagger a_c \rangle = 0.5$ and $\langle a_v^\dagger a_v \rangle = 1$, the oscillation amplitude is decreased again, cf. Fig. 4.8(c). The photon cannot be absorbed, since the excited and ground state are Pauli-blocked. Due to the Hartree-Fock terms, this Pauli-blocking is enhanced: a specific phenomenon of the many-particle interaction in semiconductor cavity-QED. In Fig. 4.8(d), the Pauli-blocking is full and cannot be enhanced. Excited and ground state are populated. The photon cannot be absorbed and no further photon can be emitted. The $g^{(2)}(t,0)$-function is constant and 0.5. The total number of electrons and holes determines the oscillation amplitude of the $g^{(2)}(t,0)$-function and of the electron and hole density. This corresponds to the impact of the photon numbers. If the photon number is high in the cavity, the oscillation amplitude of the $g^{(2)}(t,0)$-function is small. For electrons and holes, the oscillation amplitude depends on the Pauli-blocking in the system.

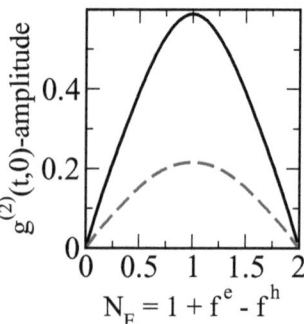

Figure 4.9: The PPCE solution for semiconductor quantum dots. The oscillation amplitude of the first two Rabi flops of the $g^{(2)}$ for PPCE is plotted over the number of electrons in the quantum dot, with $p_1 = 1.0$ (black/straight) and $p_2 = 1.0$ (red/dashed).

This is illustrated more clearly in Fig. 4.9. The number of electrons and holes in the quantum dot is plotted as a function of the oscillation amplitude maximum of the $g^{(2)}(t,0)$-function, which is directly connected to the maximum of vacuum Rabi flopping. It is seen that a maximum is achieved, if the total number equals one: $N_E = 1$, i.e. $f_e = f_h$, corresponding to the one-electron assumption. This behaviour can be attributed to the Pauli-blocking terms $f_e^n \cdot f_h^n$, because they drive the emission and are enhanced in the one electron case. In conclusion: A modified cluster expansion scheme is introduced to describe semiconductor quantum dot - wetting layer quantum optical devices in the strong coupling limit involving few photons and few dominant electronic states. This photon-probability-cluster expansion (PPCE), reproduces the JCM as a well-known benchmark in quantum electrodynamics, which the standard cluster expansion (SCE) cannot reproduce. To illustrate its strength, the new set of equations are applied to the case of vacuum Rabi flopping, where quantum fluctuations are dominant. For the QD-WL system, it is shown that the amplitude of the vacuum Rabi flopping and the maximum of $g^{(2)}(t,0)$ depend on the numbers of electrons and holes pumped into the actual quantum dot device. This is a first important prediction from the PPCE. The next step is to introduce the environment coupling to the PPCE, e.g. cavity loss and electrical pumping mechanism.

4.4 PPCE and environment coupling

Semiconductor QDs in this thesis are assumed to be self-assembled QDs, grown with the Stranski-Krastanoff method [GSB95]. The optical and electrical properties are strongly influenced by the surrounding material, the bulk material with the phonons, and the wetting layer, and or the carrier reservoir [Sti01]. The environment coupling includes dephasing processes as well as pumping processes. In the following, two important examples are given: the cavity loss, calculated within the PPCE and the electrical pumping. Hereby, the PPCE is introduced as a theoretical framework, pronouncingly able to include these coupling mechanism microscopically [SRK[+]10; RCSK09a; CRD[+]10; DMR[+]10]. For

application studies, it is highly desirable to take into account the parameter dependencies and to introduce a theoretical framework, in which the environmental coupling mechanisms can be studied and their impact on the emission properties can be predicted.
In particular, carrier inscattering are of interest for technological applications, cf. Sec. 4.4.1, emission into non-lasing modes expressed with β-factor in Sec. 4.4.2, and cavity loss in Sec. 4.4.3. This section focuses on these three environment couplings to discuss the temporal dynamic of the single photon device.

4.4.1 Electrical pumping

It is highly desirable to drive devices electrically, instead optical pumping [LST+09]. To minitiuarize the device, injection current, carrier scattering, and quantum light emission is the ideal choice [BAS+06]. In a QD, embedded in a carrier reservoir, which is embedded in the semiconductor bulk material, carrier injection can be experimentally realized [YKS+02]. This is one of the great advantages of semiconductor physics, and it is the goal to exploit the advantages and at the same time to overcome the known problems with semiconductor nanostructure based devices, such as the high dephasing due to phonons or Coulomb scattering [BLS+02].
In an electrically pumped QD structure carriers are injected into the bulk material, relax into the WL / carrier reservoir and from there into the optically active (3D confined) QD states [PRS+07]. There are two basic scattering channels: i) relaxing carriers that translate their energy to other carriers via Coulomb interaction [SCK01] and ii) carriers relax by emission of phonons [DMR+10]. The first case is important at high carrier densities, at low carrier density regime it is focused on the electron-phonon interaction, using the effective Hamiltonian approach, cf. Sec. 3.5.6. Phonons and carriers, external electrons, in the carrier reservoir are treated as bath. It is assumed that cavity mode photons and electrons are uncorrelated to the inscattering electrons and holes. As a result, the density matrix ρ of the full system can be expressed at any time as $\rho(t) \approx \rho_{sys}(t) \otimes \rho_{bath}(t)$ with a system part $\rho_{sys}(t)$ describing the quantum dots low energy electron states and the cavity mode and with $\rho_{bath}(t)$ describing all other electronic states and bosonic baths.
Therefore for any electronic states of the bath i,j and for the QD electronic states k and l, the factorization rules read $\langle |n\rangle\langle n|a_i^\dagger a_k^\dagger a_l a_j\rangle \approx \langle |n\rangle\langle n|a_k^\dagger a_l\rangle\langle a_i^\dagger a_j\rangle$. The in-scattering rates S^{in} and the out-scattering rates S^{out} for electrons and holes are calculated via this factorization and a corresponding interaction Hamiltonian between the QD and WL electronic states, including blocking and heating self-consistently [KMR+10a]. Here, the in-scattering and out-scattering is taken as a parameter. The electron and hole dynamics restricted to the electrical pump mechanism reads:

$$\partial_t f_n^e|_{pump} = S_e^{in}(p_n - f_n^e) - S_e^{out} f_n^e \qquad (4.40)$$
$$\partial_t f_n^h|_{pump} = S_h^{in}(p_n - f_n^h) - S_h^{out} f_n^h \qquad (4.41)$$

So only an index n is added to f_n and 1 is replaced by p_n compared to the SCE inscattering terms [SRK+10; CRD+10]. Note, that S^{in} and S^{out} do not depend on the photon manifold: Such a dependence will appear if the correlations with the photons is also considered for the inscattering electronic states or cross terms between the inscattering mechanisms and the electron-photon interaction. Whether the pump processes are phonon- or Coulomb assisted, the scattering rates S differ and include the specific properties of the interaction between electron and holes [DMR+10]. Furthermore, the external pumping introduces a contribution to the dephasing of the microscopic polarization:

$$\begin{aligned}\partial_t \left\langle |n+1\rangle\langle n| a_v^\dagger a_c \right\rangle |_{pump} &= -\frac{1}{2}(S_e^{in} + S_h^{in} + S_e^{out} + S_h^{out})\langle |n+1\rangle\langle n| a_v^\dagger a_c\rangle \\ &= -\gamma_p^c \langle |n+1\rangle\langle n| a_v^\dagger a_c\rangle\end{aligned} \quad (4.42)$$

due to the out scattering from the QD ground state to other states. γ_p^c is the pure dephasing contribution of the in- and out-scattering of carriers. In particular in the strong pumping regime, the scattering induced dephasing term is important.

4.4.2 β-factor

To simulate a realistic laser dynamics inside the cavity, one needs to take into account the emission into non-lasing modes [CKI94], leading to a loss of excitation of the electronic system without a contribution to the photon density. To derive this loss, an interaction Hamiltonian with the non-resonant, non-lasing modes is considered:

$$H_{\text{el-nl}} = -\hbar \sum_{i=1}^{N} M_{k_i} a_v^\dagger a_c c_{k_i}^\dagger + \text{h.c.}, \quad (4.43)$$

in which k_0 is the cavity mode and all other k_i are leading to a loss, i.e. are not strongly coupled to the QD inside the cavity [YTC00; Car99]. The completeness relation is fulfilled with $n := n_{k_0}$ and $n_{k_1} \cdots n_{k_N}$, counting the photons in the lasing and non-lasing mode, respectively:

$$\sum_n \sum_{n_{k_1}} \cdots \sum_{n_{k_N}} |n, n_{k_1} \cdots n_{k_N}\rangle\langle n, n_{k_1} \cdots n_{k_N}| = \sum_{n,\{n_k\}} |n, \{n_k\}\rangle\langle n, \{n_k\}| = \mathbb{1}. \quad (4.44)$$

With the completeness relation of the modes, it is convenient to trace out the non-lasing modes to study the laser dynamics of the system, for instance the dynamics of the photon-assisted electron and hole densities:

$$\sum_{n,\{n_k\}} |n, \{n_k\}\rangle\langle n, \{n_k\}| = \sum_n |n\rangle\langle n| \otimes \mathbb{1}_{\{n_k\}}, \quad (4.45)$$

where n counts the photons in the lasing modes and $\{n_k\}$ the photons in the non-lasing modes. The interaction with the non-lasing modes result in a decay of the electron and hole densities. Excitation is lost, but no photons in the cavity mode are created. The dynamics of the photon-density assisted electron and hole density reads:

$$\partial_t f_{n,\{n_q\}}^{e/h} \tag{4.46}$$
$$= 2\sqrt{n+1}\,\text{Im}\left[M\,\langle a_v^\dagger a_c|n+1,\{n_k\}\rangle\langle n,\{n_k\}|\rangle\right]$$
$$+ 2\sum_{i=1}^{N}\sqrt{n_{k_i}+1}\,\text{Im}\left[M_{k_i}\,\langle a_v^\dagger a_c|n,n_{k_1}\cdots n_{k_i}+1\cdots n_{k_N}\rangle\langle n,\{n_k\}|\rangle\right]$$

The non-dynamical contributions of the non-lasing modes is sumed up to study the laser dynamics in the cavity:

$$\sum_{\{n_q\}}\partial_t f_{n,\{n_q\}}^{e/h} = \partial_t f_n^{e/h} \tag{4.47}$$
$$= 2\sqrt{n+1}\,\text{Im}\left[M\,\langle a_v^\dagger a_c|n+1\rangle\langle n|\rangle\right]$$
$$+ 2\sum_{i=1}^{N}\sum_{\{n_{k_i}\}}\sqrt{n_{k_i}+1}\,\text{Im}\left[M_{k_i}\,\langle a_v^\dagger a_c|n,n_{k_1}\cdots n_{k_i}+1\cdots n_{k_N}\rangle\langle n,\{n_k\}|\rangle\right].$$

Now, the dynamics of the photon-assisted polarization of the non-lasing must be investigated, since in this equation the spontaneous emission enters the dynamics via the electron density. The process of spontaneous emission is crucial for the β-factor, here the spontaneous emission into the non-lasing mode, which is not strongly coupled to the electron transition in the QD:

$$\partial_t\langle a_v^\dagger a_c|n,n_{k_1}\cdots n_{k_i}+1\cdots n_{k_N}\rangle\langle n,\{n_k\}|\rangle \tag{4.48}$$
$$= -i\,(\omega_{cv} - \omega_{k_i} + i\gamma_p)\langle a_v^\dagger a_c|n,n_{k_1}\cdots n_{k_i}+1\cdots n_{k_N}\rangle\langle n,\{n_k\}|\rangle$$
$$- i\,M_{k_i}\sqrt{n_{k_i}+1}\,f_{n,\{n_k\}}^{e} - i\,M_{k_i}\sqrt{n_{k_i}+1}\,(p_{n,n_{k_1}\cdots n_{k_i}+1\cdots n_{k_N}} - f_{n,n_{k_1}\cdots n_{k_i}+1\cdots n_{k_N}}^{h}),$$

where the pure dephasing γ_p is included, e.g. due to longitudinal acoustical phonon interaction of the QD carriers with the bulk phonons. Non-lasing modes do not contribute to the photon density inside the cavity. In consequence, quantities such as $\langle a_v^\dagger a_v|n_k\rangle\langle n_k|\rangle = 0, \langle a_c^\dagger a_c|n_k\rangle\langle n_k|\rangle = 0$ vanish for $k \neq k_0$ and $n_k \geq 1$, and the photon probability distribution reduces to

$$\langle c_{k_i}^\dagger c_{k_i}\rangle = \sum_{n_{k_i}=0} n_{k_i}\langle |n_{k_i}\rangle\langle n_{k_i}|\rangle = 0. \tag{4.49}$$

It is valid for all time: $p_{n,n_{k_1}\cdots n_{k_i}+1\cdots n_{k_N}} = 0 = f_{n,n_{k_1}\cdots n_{k_i}+1\cdots n_{k_N}}^{h}$, since no occupation in the non-lasing modes is possible. The validity of this assumption depends on the quality factor of the cavity to support only one mode [Vah03]. No feedback of non-lasing mode photons is possible and thus, no induced processes. Therefore, the polarization dynamics can be solved in the adiabatic limit, bearing in mind that higher order photon-assisted processes and strong coupling interaction mechanism do not enter into this equation [SRK+10], and the time dependence is omitted, as well:

$$\langle a_v^\dagger a_c | n, n_{k_1} \cdots n_{k_i} + 1 \cdots n_{k_N} \rangle \langle n, \{n_k\} | \rangle = \frac{M_{k_i} \sqrt{n_{k_i}+1}}{\omega_{cv} - \omega_{k_i} + i\gamma_p} f_{n,\{n_k\}}^e \quad (4.50)$$

$$= M_{k_i} \sqrt{n_{k_i}+1} \frac{\omega_{cv} - \omega_{k_i} - i\gamma_p}{(\omega_{cv} - \omega_{k_i})^2 + \gamma_p^2} f_{n,\{n_k\}}^e.$$

Now, the result in Eq. (4.50) enters in Eq. (4.47):

$$\partial_t f_n^{e/h} \quad (4.51)$$

$$= 2\sqrt{n+1}\,\text{Im}\left[M\langle a_v^\dagger a_c | n+1\rangle\langle n|\rangle\right] - 2\gamma_p \sum_{i=1}^N \sum_{\{n_k\}} \frac{|M_{k_i}|^2 (n_{k_i}+1)}{(\omega_{cv}-\omega_{k_i})^2 + \gamma_p^2} f_{n,\{n_{k_i}\}}^e.$$

The vanishing non-lasing mode density needs to be taken into account: $n_{k_i} = 0$ and, assuming the vanishing contribution of non-lasing mode photons to the system dynamics, the sum over the non-lasing modes can be computed:

$$\partial_t f_n^{e/h} = 2\sqrt{n+1}\,\text{Im}\left[M\langle a_v^\dagger a_c | n+1\rangle\langle n|\rangle\right] - \frac{1}{\tau_{nl}} f_n^e, \quad (4.52)$$

in which the rate of the spontaneous emission into the non-lasing mode is defined as:

$$\frac{1}{\tau_{nl}} = 2\gamma_p \sum_{i=1}^N \frac{|M_{k_i}|^2}{(\omega_{cv}-\omega_{k_i})^2 + \gamma_p^2}. \quad (4.53)$$

Since β defines the fraction of the total spontaneous emission $\frac{1}{\tau_{sp}}$ into the lasing-mode $\frac{1}{\tau_l}$, one may write:

$$\beta = \frac{\frac{1}{\tau_l}}{\frac{1}{\tau_{sp}}} = \frac{\frac{1}{\tau_l}}{\frac{1}{\tau_l} + \frac{1}{\tau_{nl}}} \longrightarrow \frac{1}{\tau_{nl}} = \left(\frac{1}{\beta} - 1\right)\frac{1}{\tau_l} = \frac{1-\beta}{\beta\tau_l} = \frac{1-\beta}{\tau_{sp}}. \quad (4.54)$$

The total rate of spontaneous emission depends on the microcavity and its quality-factor, and is enhanced by the Purcell effect and can be assumed in the order of $\tau_{sp} = 50$ ps [RSL+04]. The β-factor is in the following used as an input-parameter.

The coupling to the β-factor is changed in the QD-carrier reservoir scheme, since the source term of the spontaneous emission is modified, Sec. 4.2.4. The derivation is similar, but Eq. (4.50) reads in the QD-WL case:

$$\langle a_v^\dagger a_c | n, n_{k_1} \cdots n_{k_i} + 1 \cdots n_{k_N} \rangle \langle n, \{n_k\} | \rangle \quad (4.55)$$

$$= M_{k_i} \sqrt{n_{k_i}+1} \frac{\omega_{cv} - \omega_{k_i} - i\gamma_p}{(\omega_{cv} - \omega_{k_i})^2 + \gamma_p^2} \langle a_c^\dagger a_v a_v^\dagger a_c | n, \{n_k\}\rangle\langle n, \{n_k\}|\rangle,$$

with the consequence after the modified Hartree-Fock factorization, that the electron/hole dynamics is changed to:

$$\partial_t f_n^{e/h} = -\frac{1-\beta}{\tau_{sp}} \frac{f_n^e f_n^h}{p_n} + 2\sqrt{n+1}\, \mathrm{Im}\left[M\, \langle a_v^\dagger a_c | n+1 \rangle \langle n | \rangle \right]. \quad (4.56)$$

In Eq. (4.56), the environment is included in the modified decay factor of the electron and hole densities and is a remarkable improvement to previous theoretical approaches in the few photon and few emitter dynamics, discussing semiconductor specific properties of quantum light emitter.

4.4.3 Cavity loss

To exploit the properties of the emitted quantum light, e.g. to excite another quantum system in a coupled quantum system setup [KK08; Car93], a cavity loss needs to be considered [Car99]. In particular for single-photon emission, the cavity loss parameter is of great importance, since one photon needs to be emitted at a time, before a second photon is created via induced emission processes. Focusing on repetition rate, the cavity loss is the crucial factor, e.g. how fast the system enters the steady state. Theoretically, cavity loss results from a tunnel Hamiltonian, in which dissipative photon modes q couple via the mirror with the cavity photons, cf. Fig. 4.10 [YTC00]. These photon states outside of the cavity are coupled to the cavity modes [Car99]. The coupling of the cavity modes to the external modes is decribed by an additional part in the Hamilton operator [VWW01], which reads together with the homogenous part of the external photons:

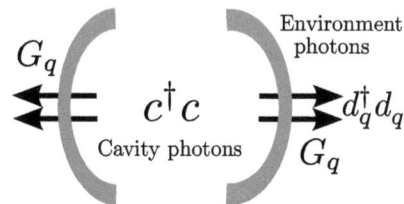

Figure 4.10: Scheme of cavity loss.

$$H_{pt-pt} = \hbar \sum_q \omega_q d_q^\dagger d_q + \hbar \sum_q G_q (c^\dagger + c)(d_q^\dagger + d_q), \quad (4.57)$$

The probability of the tunneling process from the cavity into the environment is determined by the strength of the coupling element G_q of the mode q. The cavity loss via external photon modes is described by $d_q, d_q^\dagger$. Only contributions within the rotating-wave approximation are considered. The dynamics, calculated only for the photon-photon interaction reads for an arbitrary QD operator A assisted by $|n\rangle\langle m|$ cavity photons:

$$\begin{aligned}
\partial_t \langle A|n\rangle\langle m|\rangle|_{pt-pt} &= i(n-m)\omega_0 \langle A|n\rangle\langle m|\rangle \quad (4.58)\\
&+ i\sum_q G_q \left(\sqrt{n+1}\langle A|n+1\rangle\langle m|d_q\rangle + \sqrt{n}\langle A|n-1\rangle\langle m|d_q^\dagger\rangle \right)\\
&+ i\sum_q G_q \left(\sqrt{m+1}\langle A|n\rangle\langle m+1|d_q^\dagger\rangle + \sqrt{m}\langle A|n\rangle\langle m-1|d_q\rangle \right).
\end{aligned}$$

Applying a rotating frame for convenience, for details App. 6.8, the dissipative photon mode assisted quantities are calculated, to derive the set of equations up to the second order in the tunneling matrix couling element. Following as an example is given:

$$\partial_t \langle A|n\rangle\langle m|d_q\rangle_R|_{pt-pt} \approx -iG_q \sqrt{m+1} \langle A|n\rangle\langle m+1|\rangle_R \, e^{-i(\omega_0-\omega_q)t}. \qquad (4.59)$$

Contributions proportional to $\langle d_q^{(\dagger)} d_{q'}^{(\dagger)} \rangle$ are neglected within the bath assumption, since only dissipation is taken into account [See App. 6.8 for details]. The bath photon do not couple back into the cavity. The mean photon number is too low in the bath to consider a tunnel process in the reversed direction. The equation is solved by integrating formally and applying the Markovian approximation and assuming memory effects are negligible. This is a good approximation in system-bath processes with coupling strength suggesting interaction times higher than ps-scale. The solution reads:

$$\langle A|n\rangle\langle m|d_q^\dagger\rangle_R(t)|_{pt-pt} \approx iG_q \sqrt{n+1} \langle A|n+1\rangle\langle m|\rangle_R e^{i(\omega_0-\omega_q)t} \int_0^\infty d\tau e^{-i(\omega_0-\omega_q)\tau}. \qquad (4.60)$$

If the lower limit is not extended from 0 to $-\infty$, the integral is the Heitler-Zeta function $\zeta(\omega_0 - \omega_q)$, consisting of the delta-function and of the Cauchy-principal value integral. It must be checked, to what extent the Cauchy-principal vaule integral contributes to the dynamics, normally an energy renormalization occurs, which can be included into the definition of ω_0 [Car99; SZ97]. The solution is plugged into the Eq. (4.59) to derive the cavity-loss for the general quantity $\langle A|n\rangle\langle m|\rangle$, by defining:

$$\kappa := \sum_q |G_q|^2 \, \zeta(\omega_0 - \omega_q), \qquad (4.61)$$

the PPCE formulation of the Markovian cavity loss, including energy renormalization is derived and can be expressed as:

$$\partial_t \langle |n\rangle\langle m|A\rangle|_{ph-ph} = -(m+n)\kappa \langle |n\rangle\langle m|A\rangle + 2\sqrt{m+1}\sqrt{n+1}\kappa \langle |n+1\rangle\langle m+1|A\rangle. \qquad (4.62)$$

In the PPCE formalism, the dynamics of the photon loss describes the loss and gain of photons of the different photon probabilities self-consistently. The dynamics of the photon probabilities are easier in the electron-photon interaction, but more complicated for the photon-loss, since inscattering from higher order photon probabilities need to be considered. Therefore, the cavity loss appears more complicated than in the picture of the $c^\dagger, c$-operator picture. An example is given in the App. 6.8. The photon-probability approach opens up further insight into the combined electron-photon dynamics, even in the case of the Markovian-derived cavity loss. The few-electron approach, i.e. the need to factorize in the electronic system, does not enter in this derivation of the cavity loss, since the cavity loss results solely from the photon-photon interaction.

4.4.4 Quantum dot - wetting layer laser equations

After discussing important processes, that couple the QD-WL cavity QED dynamics to the environment, the results are summarized and concluded with following equations of motion. Starting with the probability to have n photons in the system, it is obtained:

$$\partial_t p_n = -n2\kappa p_n + (n+1)2\kappa p_{n+1} \qquad (4.63)$$
$$-2\sqrt{n}\,\text{Im}\left[M\langle a_v^\dagger a_c|n\rangle\langle n-1|\rangle\right] + 2\sqrt{n+1}\,\text{Im}\left[M\langle a_v^\dagger a_c|n+1\rangle\langle n|\rangle\right],$$

including now the cavity loss process. The impact of the environment coupling on the dynamics of the transistion is larger, since electronic and photonic observables are included:

$$\partial_t \langle a_v^\dagger a_c |n+1\rangle\langle n|\rangle = \qquad (4.64)$$
$$= -i(\omega_{cv} - \omega_0 - i\gamma)\langle a_v^\dagger a_c|n+1\rangle\langle n|\rangle$$
$$- \kappa(2n+1)\langle a_v^\dagger a_c|n+1\rangle\langle n|\rangle + \kappa\sqrt{(n+1)(n+2)}\langle a_v^\dagger a_c|n+2\rangle\langle n+1|\rangle$$
$$- iM\sqrt{n+1}\left[\frac{f_n^e f_n^h}{p_n} - \frac{(p_{n+1} - f_{n+1}^h)(p_{n+1} - f_{n+1}^e)}{p_{n+1}}\right],$$

where the different pure dephasing contributions are included in γ (pump process and contributions originating from the interaction with LA phonons). The non-lasing modes do not change the strongly coupled microscopic polarization dynamics of the cavity photon assisted transition. The β-factor enters only in the dynamics of the electron and hole densities, as well as the pump mechanism:

$$\partial_t f_n^{e/h} = -\frac{1-\beta}{\tau_{sp}} \frac{f_n^e f_n^h}{p_n} + 2\sqrt{n+1}\,\text{Im}\left[M\langle a_v^\dagger a_c|n+1\rangle\langle n|\rangle\right] \qquad (4.65)$$
$$- 2\kappa\left[n f_n^{e/h} - (n+1)f_{n+1}^{e/h}\right] + S_{in}^{e/h}(p_n - f_n^{e/h}) - S_{out}^{e/h} f_n^{e/h}.$$

These equations together constitute the theoretical framework of the single photon emitter realized by a single quantum dot. Parameter dependent quantum light emission can now be studied and specific semiconductor properties be discussed. Conveniently, the full information about the quantum light is self-consistently included in the p_n distribution, which is time resolved accessible.

4.5 Electrically pumped single photon emitter

Due to the environment, the dynamics of the QD carriers is strongly damped. Pure dephasing, cavity loss and the spontaneous emission into non-lasing modes lead to dephasing and typical strong coupling signatures like Rabi oscillations are not seen. Furthermore, the pump mechanism leads to an additional dephasing and the dynamics is additionally damped via carrier relaxation and scattering from the WL to the QD states. First, the laser dynamics of an electrically driven QD is discussed, cf. Sec. 4.5.1, in different pumping regimes, from weak to strong pumping. This leads to different cavity field photon-statistics from the anti-bunching to the coherent regime. In Sec. 4.5.2, parameter dependent quantum light emission is studied, i.e. the dependence of the $g^{(2)}(t,0)$-function on the electron-photon coupling strength, the cavity loss and detuning between the cavity mode ω_0 and the QD transition energy ω_{cv}. Finally, the atomic rate equation model is compared with the semiconductor model in Sec. 4.5.3. It is shown, that enhanced Pauli-blocking in semiconductor quantum light devices is advantageous for single-photon emission.

4.5.1 Laser dynamics of an electrically driven QD

In Fig. 4.11, the dynamics of the carrier and photon related observables are depicted without an electrical pump mechanism. The QD is populated initially with an electron density of $f^e = 1$ and a hole density of $f^h = 0.7$. One photon is initially in the cavity. The photon probability distribution reads: $p_0 = 0, p_1 = 1, p_n = 0$ for $n \geq 1$. A cavity loss of $\kappa = 0.018$ meV and a pure dephasing of $\gamma_p = 0.1$ meV takes the environment coupling into account. The cavity has a β-factor of 0.9 and a Purcell enhanced spontaneous emission rate of $\tau_{sp} = 50$ ps and the QD transition frequency ω_{cv} is in resonance with the cavity mode ω_0 [RSL$^+$04]. These parameters are chosen for all calculations in this section with a electron-photon coupling strength of $M = 0.22$ meV [MIM$^+$00].
On the left of Fig. 4.11, the carrier dynamics of the QD is depicted. The electron density f^e (blue, solid line) shows oscillation as a feedback photon phenomenon of strong coupling. The electron density is driven by the microscopic polarization $\langle a_v^\dagger a_c \rangle$ (orange, dotted line). The polarization shows the oscillations of the phase filling factor or inversion $\langle a_c^\dagger a_c \rangle - \langle a_v^\dagger a_v \rangle$, including the process of spontaneous emission [SZ97]. Here, the pure dephasing attacks the polarization dynamics and stops the oscillation after 35 ps. In consequence, the Rabi oscillation of the electron density stops at the same time. The polarization is damped to zero and after the polarization dynamics comes to a stop, the electron density is damped only via the cavity loss and the spontaneous emission into non-lasing modes. The right side of Fig. 4.11 shows the dynamics of the photon observables. The photon density $\langle c^\dagger c \rangle$ (red, solid line) is one at $t = 0$ due to the initial conditions. Since the QD is initially populated, the exciton recombines and an additional photon is created. The photon density rises. Due to the cavity loss, the photon occupation is decreased before the photon density reaches 2. The photon density decays to zero in dependence on the cavity quality factor. The oscillation of the photon density stops after 35 ps, also. The intensity-intensity correlation $g^{(2)}(t,0)$-function (green, dotted line) shows longer oscillation due to the cavity loss mechanism. The oscillation is expected to peak at around 0.5, since single photon interact in the cavity. Anti-bunching is visible. For the time scale of the decaying electron density, single photons are emitted out of the cavity until the whole dynamics stops. Without pumping mechanism, the stationary state is zero for the observables and for the polarization. However, if the system is pumped, a non-trivial stationary state is reached with values inequal to zero.
Now, a stationary electrical pump mechanism is considered: $S_{e/h}^{in/out} \neq 0$. Initially, the QD is unoccu-

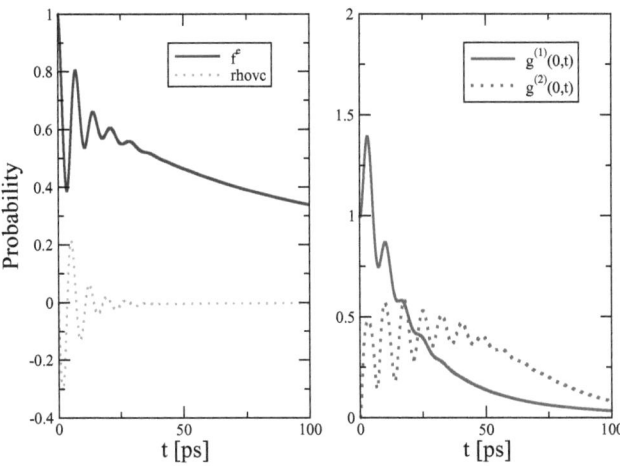

Figure 4.11: QD as a single photon emitter, using the parameters $M = 0.22$ meV, $\gamma_p = 0.1$ meV and $\kappa = 0.018$ meV without electrical pumping. Dynamics show Rabi oscillation.

pied. A typical, low carrier density regime ($8 \times 10^9 (cm)^{-2}$) value of thermal occupation probability is around $f^e \approx 0.001$ at 70 K. Therefore, the initial conditions are chosen with $f^{e/h}(0) = 0$ and no photons are assumed in the cavity: $p_0(0) = 1, p_n(0) = 0$ for $n \geq 1$. In the following, holes and electrons are equally pumped: $S_{in}^e = S_{in}^h = S_{in}$ and $S_{out}^e = S_{out}^h = S_{out}$, and three pump regimes are investigated: weak, transient and strong pumping strength. These regimes correspond to a transition from single-photon emission (weak) to lasing well above threshold (strong).

In Fig. 4.12(left), a weak electrical pumping drives the QD dynamics: $S_{in} = 4 ns^{-1}$ and the out-scattering is $S_{out} = 0.1 S_{in}$. With increasing pump duration, an electron and a hole density is built up in the QD (solid, blue line). After 10 ps, electrons and holes start to recombine. The photon density (dotted, green line) rises. Emission processes are slowed down by the dephasing in the system and the cavity loss leads to a loss of photons as well. After 100 ps, the $g^{(2)}(t,0)$-function reaches a stationary value of approximately 0.3, within the single-photon limit. The single photon emitter now operates in this stationary state. Single photons are emitted out of the cavity. Excitons are created and recombine in balance. Within the PPCE approach, the full information about the quantized light field is easily obtained via the photon probability distribution [SRK+10; RCSK09a; MW95]. In Fig. 4.12(right), the photon probability distribution in the stationary limit is depicted. A Fock distribution is visible. Only two photon probabilities are inequal to zero. Rabi oscillations lead to an increase of p_1. Prepared in a pure Fock state, the cavity field oscillates only between two photon probabilities, if only one emitter, e.g. QD, is placed inside the cavity. Here, the stationary value has a maximum of p_0, but enough probability p_1 to actually emit a photon from the cavity. A single photon source needs a small $g^{(2)}(t,0)$-function with a high probability of a photon emission: $\langle c^\dagger c \rangle$. In Fig. 4.12, both conditions are fulfilled. The photon probability resolved dynamics and results are important to understand quantum

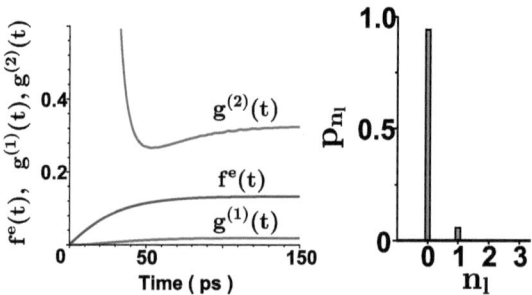

Figure 4.12: QD as a single photon emitter, using the parameters $M = 0.22$ meV, $\gamma_p = 0.1$ meV and $\kappa = 0.018$ meV, weakly pumped: $S_{in} = 4\text{ns}^{-1}$ and the out-scattering is $S_{out} = 0.1 S_{in}$.

light correlation on a fundamental level, e.g. the importance of p_2 to find a strongly anti-bunched cavity field. Physically, this corresponds to negligible induced emission in the cavity. The electron and hole density is not built up fastly enough. An induced emission process cannot take place. The next interesting regime is the transient regime for medium pumping.

In the transient regime, the pump mechanism is fast enough to repopulate the QD before the cavity photon is reabsorbed and creates an exciton. In Fig. 4.13(left), the observables of the system is shifted to a higher stationary value due to the stronger pumping: $S_{in} = 0.1\text{ps}^{-1}$ and the out-scattering is $S_{out} = 0.1 S_{in}$. The electron density (solid, blue line) is built up faster. The inscattering is stronger and the electron density saturates on a higher stationary value. Due to the higher electron and hole density, more excitons can recombine on a shorter time scale and a higher photon number is reached with a stationary value of approximately 0.1. The stationary value is reached faster due to the faster inscattering processes and the additional dephasing, which is introduced in the photon-assisted polarization. The $g^{(2)}(t, 0)$-function saturates at a value of 0.55, above the single-photon level. To understand the transient regime in more detail, the photon probability distribution is plotted on the right hand side

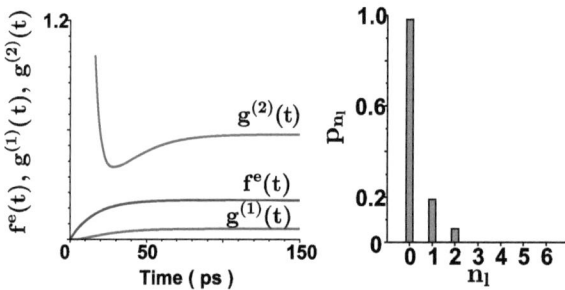

Figure 4.13: QD as a single photon emitter, using the parameters $M = 0.22$ meV, $\gamma_p = 0.1$ meV and $\kappa = 0.018$ meV, pumped in the transition regime.

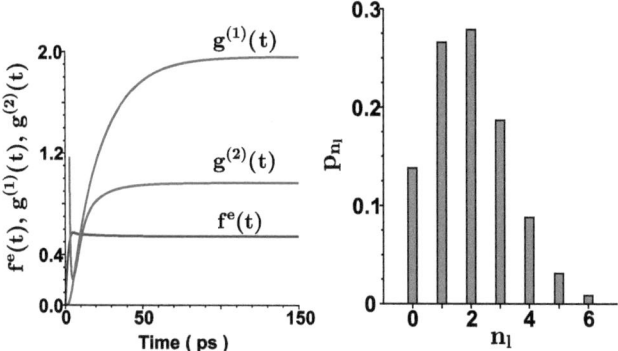

Figure 4.14: QD as a single photon emitter, using the parameters $M = 0.22$ meV, $\gamma_p = 0.1$ meV and $\kappa = 0.018$ meV. The electrical pumping is strong. The QD starts to enter the lasing regime.

of Fig. 4.13. Here, clearly visible, is the condition of a single photon emitter not fulfilled. It is highly probable to find a second photon in the cavity with around 7%. If the goal is a single QD laser, the opposite is to provide, or a stronger pumping.

A lot of effort is aiming at a single QD laser, a very efficient laser device, operating above threshold to provide a coherent quantum light emission [MKZ+00]. Conditions for a single QD laser is a high-Q cavity with a strong electron-light coupling, exceeding the losses to provide induced emission processes. A β-factor close to 1 is necessary, for maximally quantum exploit efficiency, as well as a fast pump mechanism, e.g. electrical pump mechanism via a carrier reservoir [KMR+10b]. In Fig. 4.14(left), the dynamics of a single QD laser is plotted. The increased pump rates $S_{in} = 0.33$ps^{-1} and the out-scattering $S_{out} = 0.1 S_{in}$ enforces the electron density (solid, blue line) to saturated very fast within only 5 ps. After 10 ps, electron and photon density cross and the photon density saturates at a stationary value of approximately 2. On approximately the same time scale, the $g^{(2)}(t, 0)$-function reaches the stationary value close to 1, indicating a coherent cavity light field.

Specific for this pumping regime is the order, in which the observables enter into the stationary regime. For weak and medium pumping, the photon density saturates before the $g^{(2)}(t, 0)$-function. But in the strong regime, the $g^{(2)}(t, 0)$-function saturates fast and before the photon density at 60 ps. Coherence is reached and cannot be more enforced as it is at that point of pumping. In this regime, the Hartree-Fock approximation begins to become problematic. For high carrier densities, higher order correlation between electrons and holes need to be taken into account. For much higher pumping rates beyond $S_{in} = 0.33$ps^{-1}, the calculation need to be cross checked with higher order calculations, since the density grow to higher values than 0.5 and correlations cannot be neglected. Additionally, an adiabatic switch-on of the pumping mechanism need to be implemented for much higher pumping rates, since for very strong pumping rates, the instantaneous pumping of the carrier states in the QD may lead to numerical artefacts. It does not correspond to the slowly increased injection current into the bulk material, or the tunneling of carriers directly into the carrier reservoir for the QD. In Fig. 4.14(right), the photon probability distribution is depicted. A Poisson distribution is visible with the maximum at 2 photons, the stationary value, reached in the time dynamics. Up to 6 photon become probable in the cavity, which converts the initially Fock like squeezed cavity field into a Poisson-distributed coherent field with a fluctuation around the mean photon number of $\sqrt{\bar{n}}$, leading to laser emission.

Note that in the plots above Fig. 4.12-4.14, signatures of Rabi oscillation are not visible. Since the QD is initially unpopulated and the losses are too strong, the electron density does not start to oscillate. The theoretical approach is used in the next section to investigate parameter dependent photon-statistics of the emitted quantum light.

4.5.2 Parameter studies of a single QD laser device

For device optimization, parameter studies are a valuable tool to frame efficient operating points. In the following, the photon-statistics is calculated for different parameter set. A chosen parameter, e.g. the cavity loss, is varied, whereas the other parameters are held constant, to investigate the interplay between the different physical processes of gain, emission, absorption, dephasing and damping. The system is started and evaluated until a stationary state is reached. The corresponding values are plotted for the varied parameter. Since the time the stationary state is reached and the values differ strongly, the order of equations of motion need to be adjusted to the chosen parameter set. This is done by investigating the photon probabilities $\langle |n\rangle\langle n| \rangle$ to find an N_{max}, so that $p_n = 0$ for $n \geq N_{max}$. This N_{max} depends on the parameter set. However, for every given initial conditions and parameter an N_{max} can be found, bearing in mind that, this N_{max} is dynamically located and used, controlling the validity of the calculation. Note, the theoretical results provides mean probabilities. So the quantities presented here, represent mean values after enough repetitions of the experiments. Though data of single experiments can not be analysed with this methods, in this case quantum jump theories are of particular importance.

The electron-light coupling strength M is crucial, since the time scale of the emission and absorption processes depends on it. The QD interacts with the cavity photons via M. The coupling matrix element includes the dipole moment of the QD between valence and conduction band state, as well as the photon mode and the Purcell factor, determined by the qualtiy factor of the microcavity [YTC00]. Although κ depends also on the cavity parameter, the electron - light coupling is independently varied from κ, considering different dipole moments of the QD in a comparable cavity environment. The dipole moments depends on the size, on the geometry and of the material of the QD, given the respective wave functions [Sti01]. In Fig. 4.15, the $g^{(2)}(t,0)$-function (red line), the photon density (green line) and the electron density (blue line) are plotted for different electron-light coupling strength. A cavity loss of $\kappa = 1/(4.3\text{ps})$, a pure dephasing of $\gamma_p = 1/(10\text{ps})$ are chosen and scattering rates of $S^e_{in} = 1/(4\text{ps})$, $S^e_{out} = 1/(40\text{ps})$, $S^h_{in} = 1/(25\text{ps})$, $S^h_{out} = 1/(8.3\text{ps})$. A $\beta = 1$ is assumed. This parameter set determines the range of values for M, which result in a change of the stationary value. Beyond

4 QUANTUM DOT - WETTING LAYER CAVITY QUANTUM ELECTRODYNAMICS

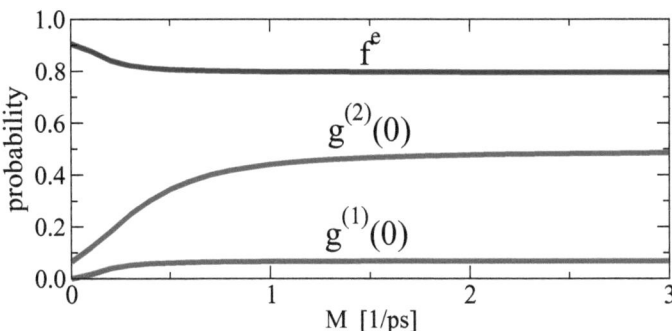

Figure 4.15: The single photon emitter in steady state condition over different values of electron-photon coupling M. Using the parameters $\kappa = 1/(4.3\text{ps})$, $\gamma_p = 1/(10\text{ps})$ and the scattering rates $S_{in}^e = 1/(4\text{ps})$, $S_{out}^e = 1/(40\text{ps})$, $S_{in}^h = 1/(25\text{ps})$, $S_{out}^h = 1/(8.3\text{ps})$, and $\beta = 1$.

this range, M can be varied without changing the stationary state value, since other processes prevent an efficient photon emission, e.g. the chosen in- and out-scattering. If the inscattering of electrons is much slower than the electron-photon coupling strength, the in-scattering is the crucial parameter and a higher value of the electron-photon interaction does not change the quantum light emission. This is visible in Fig. 4.15. After a value of $M = 0.5$ ps^{-1}, the photon density and the electron density do not change much for higher values for M. The electron-photon coupling strength exceeds the inscattering rate too much and loses the impact on the stationary value. Only the $g^{(2)}(t,0)$-function is rising slightly afterwards due to the fact, that higher correlations in the photon field are driven more efficiently with a higher M via spontaneous emission processes. The electron density f^e (blue line) is decreased for stronger electron-photon coupling: excitons in the QD recombine faster with an increased M. In consequence, the photon density (green line) rises with increasing M, until the cavity loss comes into play and stops the continous increment of photons in the cavity, as well as the inscattering rates of electron and holes. For this set of parameter, the $g^{(2)}(t,0)$-function is increasing, but remains in the single-photon limit, i.e. below a value of 0.5. With given in-scattering and losses, a single QD laser could not be realized, even with a very high dipole moment. Other processes prevent the increment of $g^{(2)}(t,0)$-function to 1.

Another important parameter is the cavity loss with $\kappa = \omega_0/Q$, defining Q as the quality factor of the cavity, cf. 3.1.1. On the one hand, it is highly desirable to have an efficient out-coupling of photons from the cavity, to produce a fast triggered-single-photon source [LO05]. On the other hand, a very high-Q cavity is desirable with strong electron-light coupling, a high Purcell enhancement to create photons in the desired mode. Hence, one needs a low and a high cavity loss at the same time. The impact of the cavity loss parameter is depicted in Fig. 4.16. Obviously, for a high κ, the photon density (green line) is decreased fast and is vanishingly small for κ beyond 1 ps^{-1}. Also, the $g^{(2)}(0)$-function (red line) is decreased due to a damping of higher order correlation in the photon field: the higher the photon correlation, the higher the losses, cf. Eq. (4.63). Although the photon-statistics of the cavity

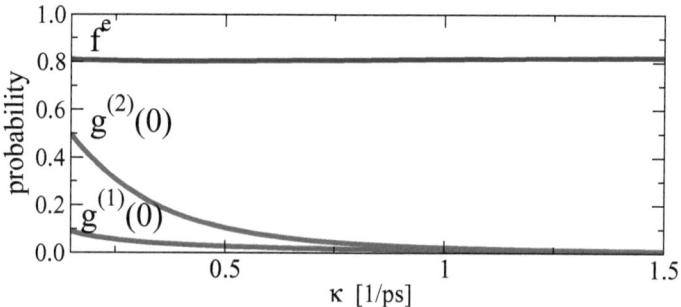

Figure 4.16: The single photon emitter in steady state condition over different values of cavity loss κ. Using the parameters $M = 1/(2\text{ps})$, $\gamma_p = 1/(10\text{ps})$ and the scattering rates $S^e_{in} = 1/(4\text{ps})$, $S^e_{out} = 1/(40\text{ps})$, $S^h_{in} = 1/(25\text{ps})$, $S^h_{out} = 1/(8.3\text{ps})$, and β is assumed to be 1.

field remains in the anti-bunching regime, the mean photon number produced by the single photon device is for high κ too small, to operate in an efficient way in terms of a single-photon device. Since κ does not increase the $g^{(2)}(t,0)$-function, or changes the photon-statistics, a very small κ is advantageous for single photon emission on demand. The $g^{(2)}(t,0)$-function decreases only as a result of the vanishing photons in the system. A quantum correlation effect is not visible and is not expected. The electron density (blue line) is not changed at all. A small increase of the stationary value of the electron density is still visible and can be explained with an additional damping of the photon-assisted densities. For high values of κ, the emission process takes longer and the electron density f^e can be filled slightly longer with the result of an increment in the stationary value of the electron density. Concluding the parameter studies, the photon density and the $g^{(2)}(t,0)$-function are investigated for different pump rates and for different detunings Δ between the QD transition frequency and the cavity mode in units of ps^{-1}. The β-factor is assumed to be 0.9. Obviously, the detuning enforces a transition from the strong coupling into the weak regime. In Fig. 4.17(left), the stationary value of the photon density $g^{(1)}(0)$ for different detuning and increasing inscattering is plotted. For a detuning $\Delta = 1$ ps^{-1} (blue line), in order of magnitude of the electron-photon coupling strength $M = 0.5$ ps^{-1}, the mean photon number depends approximately linearly on the pump rate. With increasing pump rate, the mean photon number rises also, outrivaling the cavity loss. For a stronger detuning (red and orange line) with $\Delta = 2$ ps^{-1} and $\Delta = 3$ ps^{-1}, the detuning effect is stronger than the electron-photon coupling and the mean photon number increases also, but slower. For an increasing detuning, the emission of photons is less probable, until the emission stops ultimately, if the detuning is too big and no further feeding mechanisms are considered [MKH08; Hoh10; WVT+09; TS10]. On the right hand side of Fig. 4.17, the $g^{(2)}(t,0)$-function in dependence on the pump rate and on the detuning is plotted. In contrast to the photon density, the detuning has a surprising impact on the $g^{(2)}(t,0)$-function. For a small detuning (blue line), the $g^{(2)}(t,0)$-function rises and saturates at a stationary value of approximately 1. The QD-carrier reservoir system operates in the laser limit well above threshold. A Poisson-distribution is obtained. But for a higher detuning a remarkable effect is visible. The $g^{(2)}(t,0)$-function

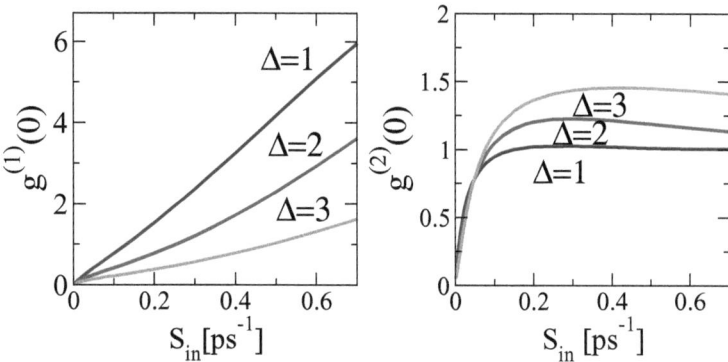

Figure 4.17: The single photon emitter in steady state condition over different values of the inscattering rate S_{in} and for different detuning Δ between the cavity mode and the QD transition energy in units of ps^{-1}. Using the parameters $M = 1/(2\text{ps})$, $\gamma_p = 1/(10\text{ps})$ and β is assumed to be 0.9.

does not progress to 1, but overshoots into the bunching regime, reaches a maximum value and progresses then back to 1 for high pump rates. This effect is widely discussed in recent experimental and theoretical work and is attributed to background noise [NKI+10], to a biexciton contribution[RGGJ10] or, like in this case, to a detuning of the QD-cavity resonance. In the Hartree-Fock limit, higher pump rates become problematic. The unfactorized correlation between electron-and holes must be negligible. Here, the investigation is focused on the single-photon limit, not on the laser limit. To investigate the laser dynamics of a single QD laser in the high pumping limit, it is necessary to go beyond the Hartree-Fock factorization in the similar manner like the SCE. The possibility of the PPCE to include correlation of higher order is one of its strength and is in preparation. Remarkable, the bunching transition dynamics is observable within the Hartree-Fock limit and within the given parameter set.

4.5.3 Atom and QD-WL rate equations in the single-photon limit

The main difference between an atomic model and a QD-carrier reservoir (WL) model stems from the validity of the one-electron assumption (OEA), which is typically valid in atomic system. In atomic systems electron and holes are correlated, e.g. via $\langle a_c^\dagger a_c \rangle + \langle a_v^\dagger a_v \rangle = 1$ [BY99]. In the atomic limit, the Hartree-Fock is not necessary. Quantities like $\langle a_c^\dagger a_v^\dagger a_v a_c \rangle$ vanish and the spontaneous emission is only driven by the electron density $f^e = \langle a_c^\dagger a_c \rangle$. In consequence, the photon-assisted polarization is strongly modified in comparison to Eq. (4.64) and reads:

$$\partial_t \langle a_v^\dagger a_c | n+1 \rangle \langle n | \rangle = (-i\Delta - \gamma_p) \langle a_v^\dagger a_c | n+1 \rangle \langle n | \rangle - iM\sqrt{n+1}(f_n^e - p_{n+1} + f_{n+1}^h). \tag{4.66}$$

The laser rate equations are derived by assuming the polarization dynamics to be stationary on the emission process time scale. Eq. (4.66) is solved in the adiabatic limit. The time-derivative is assumed to be zero. With the given equation for the polarization, using the one-electron assumption and apply-

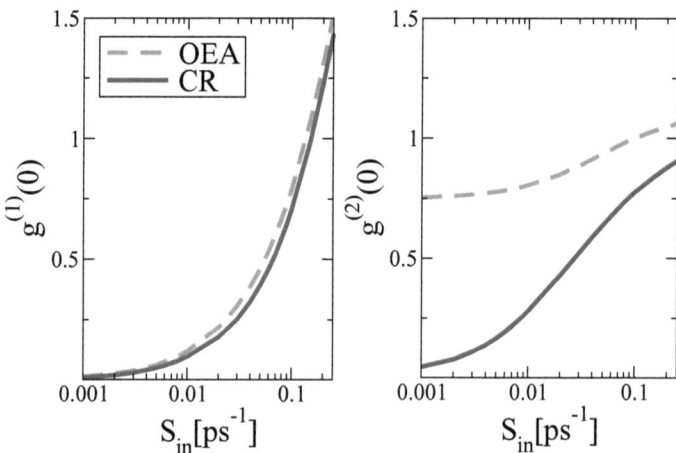

Figure 4.18: Photon density $g^{(1)}(0)$ and $g^{(2)}(t,0)$-function in a steady state condition over different values of the inscattering rate S_{in} in an atomic (OEA) and in a QD-carrier reservoir system (CR): $\Delta = 0.05$ ps^{-1}, $\kappa = 0.03$ ps^{-1} $M = 1/(5\text{ps})$, $\gamma_p = 0.05$ ps^{-1} and β is assumed to be 0.98 and $S_{out} = 0.1\, S_{in}$.

ing second order Born-approximation: $f_n^e \approx f^e p_n$, the PPCE equations Eq. (4.63) - Eq. (4.65) reduce to the well-known laser rate equation for the photon density [Hak94]:

$$\partial_t \langle c^\dagger c \rangle = -2\kappa \langle c^\dagger c \rangle + \frac{\beta}{\tau_{sp}} \left[\langle c^\dagger c \rangle + 1 \right] f^e \qquad (4.67)$$

with $\langle c^\dagger c \rangle = \sum_n n\, p_n$ and to close the set of equations, one obtains for the electron dynamics with the expression of the microscopic polarization and in second order Born:

$$\partial_t f^e = -\frac{1}{\tau_{sp}} f^e - \frac{\beta}{\tau_{sp}} f^e \langle c^\dagger c \rangle + S_{in}(1 - f^e) - S_{out} f^e. \qquad (4.68)$$

Given a constant pump rate, dephasing and cavity losses, the observable enter eventually in the stationary regime, in which the photon density and the $g^{(2)}(t,0)$-function can be compared. In Fig. 4.18, the stationary values of the photon density $g^{(1)}(0)$ (left) and the intensity-intensity correlaton $g^{(2)}(t,0)$-function (right) is plotted for different in- and outscattering $S_{in/out}$. The parameter set is given in the caption of Fig. 4.18. The cavity dynamics are evaluated for a atomic system (orange, dashed line, OEA) and for a QD-carrier reservoir system (green, solid line,CR). Comparing the stationary values of the photon density in Fig. 4.18(left), between the atomic and QD-CR-system, the difference is small. With increasing pump rate, the mean photon number rises gradually. The higher the pump rate, the faster the mean photon number rises. The atomic system shows a slightly faster output of photons from a pump rate of $S_{in} = 0.005$ ps^{-1} until $S_{in} = 0.2$ ps^{-1}. At this value, both output curves start to converge again. However, this small difference indicates a remarkable feature of the QD-CR system, concerning single-photon emission on demand. Intensity-intensity correlation are surpressed without the lack of photons. The photon density is almost the same, but not the $g^{(2)}(t,0)$-function, cf. Fig. 4.18(right). For the atomic system and with given parameters, the $g^{(2)}(t,0)$-function starts in the anti-bunching regime at a value of 0.75 and rises gradually with increasing pump rate to the coherent regime around 1. This indicates, the transition into the laser regime. However, with a similar output curve, the QD-CR system shows a wide range of strongly anti-bunched quantum light emission. Even for a pump rate of $S_{in} = 0.01$ ps^{-1}, the device operates in the single-photon regime, below a value of 0.5 due to the fact, that the semiconductor specific property, uncorrelated electron and hole densities, lead to an enhanced Pauli-blocking, which in turn leads to a decreased emission rate into the laser mode. This enhanced Pauli-blocking originates from the spontaneous emission term: $f_n^e f_n^h / p_n$, which is in the atomic case only proportional to f^e. In typical experimental data [NKI+10], the value of the $g^{(2)}(t,0)$-function in a steady state condition is much lower than theoretical calculations predict. Those numerical simulations are based on the one-electron assumption and do not include the enhanced Pauli-blocking due to the presence of the wetting layer. Typical experimental values of the $g^{(2)}(t,0)$-function are well-below 0.5 and thus in the single-photon limit. Within the PPCE, this strong anti-bunching is enforced by the modified spontaneous emission term, cf. Fig. 4.18 and agrees better with experimental data [NKI+10].

Being disadvantageous in other regimes (lasing), the single photon emission is enhanced in this case and only modelled correctly within the PPCE-approach. This proves the importance of microscopical models, which take into account carrier-carrier scattering to investigate interesting operating conditions for technological applications, and to reveal advantageous properties within a semiconductor environment.

5 Conclusion and outlook

This thesis focuses on the single-photon regime of the semiconductor quantum dot-cavity quantum electrodynamics (QD-CQED) beyond the Jaynes-Cummings model, including non-Markovian contributions and many-particle interactions such as electron-phonon coupling or enhanced Pauli-blocking on a microscopic level. Two theoretical frameworks are introduced to provide a simulation tool for parameter studies of single-photon devices: (i) the mathematical induction model solves the QD-CQED including the electron-LO phonon interaction up to an arbitrary accuracy in case of a fixed number of electrons in the quantum dot; and (ii) in the presence of a carrier reservoir, the photon-probability cluster expansion (PPCE) is devoloped and introduces a modified Hartree-Fock factorization.

Within the mathematical induction model, the conversion of a thermal into a non-classical cavity field with sub-Poissonian statistics has been shown and LO-phonon assisted cavity feeding with additional anti-crossing signatures is discussed. Present anharmonicities in the Rabi oscillations is attributed to the mixing of Rabi frequencies for elevated temperatures. The modified Rabi frequency is expressed using the Huang-Rhys factor. An analytical derivation with amplitude modulation is a next step to simplify the LO-phonon QD-CQED via an effective Hamiltonian approach.

The biexciton cascade is calculated in the strong and weak coupling regime. The polarization entanglement of the generated photons is theoretically investigated. Via an effective multi-phonon Hamiltonian approach, a temperature dependent analysis of the degree of entanglement has been performed. For elevated temperatures ($T \geq 120$ K) in GaAs, the polarization entanglement vanishes due to the presence of the wetting layer. Further investigations, in particular via the induction method in the strong coupling regime and of LO-phonon coupling to the QD excitons may reveal other probabilities for an enhanced degree of entanglement. For example, the LO-phonon sidebands may enforce an erasing of the which-path information, similiar to the modulation of the exciton and biexciton energies via external applied electrical fields.

Within the PPCE, an electrically driven single-photon emitter has been studied, including enhanced Pauli-blocking in a modified spontaneous emission source term. This many-particle specific Pauli-blocking is advantageous for generating single photons in a wide range of the electrical pump strength in comparison to atom-based single photon emitter devices. To calculate a single-QD laser operating well above threshold, the investigation needs to go beyond the introduced modified Hartree-Fock limit. As a next step, correction terms will be included into the calculation to study threshold behavior in the $g^{(2)}(t,0)$-function. Additionally, the electron-phonon interaction has to be included to motivate the phenomologically introduced pure dephasing microscopically.

5 Conclusion and outlook

6 Appendix

6.1 Numerical parameters

Parameter	Symbol	Value
electron mass	m_0	$5.6856800\ fs^2eVnm^{-2}$
electron effective mass	m_e	$0.043m_0$ [RÖ2]
hole effective mass	m_h	$0.450m_0$ [RÖ2]
LO phonon energy	$\hbar\omega_{LO}$	36.4 meV [RÖ2]
QD band gap	$\hbar\omega_{cv}$	1.5 eV
Electron confinement energy	$\hbar\omega_{el}$	50 meV
Hole confinement energy	$\hbar\omega_h$	25 meV
QD diameter	$l = \sqrt{\frac{\hbar^2 \ln 2}{m_e^* m_0 \hbar\omega_{el}}}$	15,77 nm
High frequency dielectric constant	ϵ_∞	10.9
Static dielectric constant	ϵ_{stat}	12.53

6.2 Biexciton cascade in the strong coupling regime: Equations of motions

Phenomenologically, the cavity loss is introduced κ, cf. Sec. 3.1.2. Since the two-electron assumption holds, the biexciton couples only to energetic lower states, such as the intermediate exciton states via the corresponding transitions:

$$\partial_t \langle X_H^\dagger B H^{m,n} V^{p,q} \rangle \tag{6.1}$$
$$= i\left[(m-n)\omega_H^0 + (p-q)\omega_V^0 + \omega_H - \omega_B + i\kappa(m+n+p+q)\right]\langle X_H^\dagger B H^{m,n} V^{p,q} \rangle$$
$$- iM\langle G^\dagger B H^{m+1,n} V^{p,q} \rangle - iM\,m\langle B^\dagger B H^{m-1,n} V^{p,q} \rangle$$
$$- iM\langle B^\dagger B H^{m,n+1} V^{p,q} \rangle + iM\langle X_H^\dagger X_H H^{m,n+1} V^{p,q} \rangle - iM\langle X_H^\dagger X_V H^{m,n+1} V^{p,q+1} \rangle$$

$$\partial_t \langle X_V^\dagger B H^{m,n} V^{p,q} \rangle \tag{6.2}$$
$$= i\left[(m-n)\omega_H^0 + (p-q)\omega_V^0 + \omega_V - \omega_B + i\kappa(m+n+p+q)\right]\langle X_V^\dagger B H^{m,n} V^{p,q} \rangle$$
$$- iM\langle G^\dagger B H^{m,n} V^{p+1,q} \rangle + iM\,p\langle B^\dagger B H^{m,n} V^{p-1,q} \rangle$$
$$+ iM\langle B^\dagger B H^{m,n} V^{p,q+1} \rangle + iM\langle X_V^\dagger X_H H^{m,n+1} V^{p,q} \rangle - iM\langle X_V^\dagger X_V H^{m,n} V^{p,q+1} \rangle$$

6 Appendix

In the photon-assisted exciton-biexciton transition is driven via the spontaneous emission due to the biexciton relaxation. Either a horizontal or vertical photon is emitted. Note, the frequency $\omega_0^{H/V}$ originates from the cavity frequency and must not be in resonance with transition energy $\omega_{H/V} - \omega_B$. The spontaneous emission process is crucial in the weak coupling regime. In the strong coupled dynamics, the exciton-biexciton transition is also driven by the photon-assisted exciton density:

$$\partial_t \langle X_H^\dagger X_H H^{m,n} V^{p,q} \rangle \tag{6.3}$$
$$= i\left[(m-n)\omega_H^0 + (p-q)\omega_V^0 + i\kappa(m+n+p+q)\right]\langle X_H^\dagger X_H H^{m,n} V^{p,q} \rangle$$
$$- iM\langle G^\dagger X_H H^{m+1,n} V^{p,q} \rangle - iM\, m\, \langle B^\dagger X_H H^{m-1,n} V^{p,q} \rangle - iM\langle B^\dagger X_H H^{m,n+1} V^{p,q} \rangle$$
$$+ iM\langle X_H^\dagger G H^{m,n+1} V^{p,q} \rangle + iM\, n\, \langle X_H^\dagger B H^{m,n-1} V^{p,q} \rangle + iM\langle X_H^\dagger B H^{m+1,n} V^{p,q} \rangle$$

$$\partial_t \langle X_V^\dagger X_V H^{m,n} V^{p,q} \rangle \tag{6.4}$$
$$= i\left[(m-n)\omega_H^0 + (p-q)\omega_V^0 + i\kappa(m+n+p+q)\right]\langle X_V^\dagger X_V H^{m,n} V^{p,q} \rangle$$
$$- iM\langle G^\dagger X_V V^{p+1,q} \rangle + iM\, p\, \langle B^\dagger X_V H^{m,n} V^{p-1,q} \rangle + iM\langle B^\dagger X_V V^{p,q+1} \rangle$$
$$+ iM\langle G^\dagger X_V H^{m,n} V^{p,q+1} \rangle - iM\, q\, \langle X_V^\dagger B H^{m,n} V^{p,q-1} \rangle - iM\langle X_V^\dagger B H^{m,n} V^{p+1,q} \rangle.$$

The photon-assisted intermediate exciton densities couple to the higher electronic levels via the transition to the biexciton and to the lower electronic level via the transition to the ground state. Due to spontaneous emission processes, the exciton densities are driven stronger by the biexciton transition than by the ground state transition. The exciton-exciton transition $(X_H^\dagger X_V)$ has a significant impact on the exciton dynamics, if the two intermediate exciton levels are energetically close. This exchange of excitation energy is of great importance for the degree of entanglement, since in this quantity the relaxation paths are crossing over.

$$\partial_t \langle X_V^\dagger X_H H^{m,n} V^{p,q} \rangle \tag{6.5}$$
$$= i\left[(m-n)\omega_H^0 + (p-q)\omega_V^0 + \omega_V - \omega_H + i\kappa(m+n+p+q)\right]\langle X_V^\dagger X_H H^{m,n} V^{p,q} \rangle$$
$$- iM\langle G^\dagger X_H H^{m,n} V^{p+1,q} \rangle + iM\, p\, \langle B^\dagger X_H H^{m,n} V^{p-1,q} \rangle + iM\langle B^\dagger X_H H^{m,n} V^{p,q+1} \rangle$$
$$+ iM\langle X_V^\dagger G H^{m,n+1} V^{p,q} \rangle + iM\, n\, \langle X_V^\dagger B H^{m,n-1} V^{p,q} \rangle + iM\langle X_V^\dagger B H^{m+1,n} V^{p,q} \rangle$$

The smaller the fine structure splitting, the more is this exciton-exciton transition density like and is driven strongly by the photon-assisted polarization. Due to this quantity, it is possible that vertical polarized photons are emitted in a biexciton cascade, which is driven only by horizontal polarized photons. The horizontal polarized photon density can be converted completely into vertical polarized

6 Appendix

photons, if the energy splitting between the intermediate exciton levels is smaller than the cavity loss, i.e. some μeV. The conversion is driven by the ground to exciton state transitions:

$$\partial_t \langle G^\dagger X_H H^{m,n} V^{p,q} \rangle \tag{6.6}$$
$$= i\left[(m-n)\omega_H^0 + (p-q)\omega_V^0 - \omega_H + i\kappa(m+n+p+q)\right]\langle G^\dagger X_H H^{m,n} V^{p,q}\rangle$$
$$- iM\, m\, \langle X_H^\dagger X_H H^{m-1,n} V^{p,q}\rangle - iM\langle X_H^\dagger X_H H^{m,n+1} V^{p,q}\rangle$$
$$- iM\, p\, \langle X_V^\dagger X_H H^{m,n} V^{p-1,q}\rangle + iM\langle G^\dagger GH^{m,n} V^{p,q}\rangle + iM\, n\, \langle G^\dagger BH^{m,n-1} V^{p,q}\rangle$$
$$- iM\langle X_V^\dagger X_H H^{m,n} V^{p,q+1}\rangle + iM\langle G^\dagger BH^{m+1,n} V^{p,q}\rangle$$

$$\partial_t \langle G^\dagger X_V H^{m,n} V^{p,q}\rangle \tag{6.7}$$
$$= i\left[(m-n)\omega_H^0 + (p-q)\omega_V^0 - \omega_V + i\kappa(m+n+p+q)\right]\langle G^\dagger X_V H^{m,n} V^{p,q}\rangle$$
$$- iM\langle X_H^\dagger X_V H^{m,n+1} V^{p,q}\rangle - iM\, p\, \langle X_V^\dagger X_V H^{m,n} V^{p-1,q}\rangle$$
$$- iM\, m\, \langle X_H^\dagger X_V H^{m-1,n} V^{p,q}\rangle - iM\langle X_V^\dagger X_V H^{m,n} V^{p,q+1}\rangle + iM\langle G^\dagger GH^{m,n} V^{p,q+1}\rangle$$
$$- iM\, q\, \langle G^\dagger BH^{m,n} V^{p,q-1}\rangle - iM\langle G^\dagger BH^{m,n} V^{p+1,q}\rangle$$

The ground to exciton transition is formally equivalent to the exciton to biexciton transition. It includes spontaneous emission processes, induced absorption and emission via the phase filling factor, or photon-assisted inversion [SZ97]. Two new quantities appear, the photon-assisted ground state density and the ground to biexciton state transition ($\langle G^\dagger B\rangle$), which is another crucial and characteristic quantity in the four-level exciton dynamics.

$$\partial_t \langle G^\dagger BH^{m,n} V^{p,q}\rangle \tag{6.8}$$
$$= i\left[(m-n)\omega_H^0 + (p-q)\omega_V^0 - \omega_B + i\kappa(m+n+p+q)\right]\langle G^\dagger BH^{m,n} V^{p,q}\rangle$$
$$- iM\, m\langle X_H^\dagger BH^{m-1,n} V^{p,q}\rangle - iM\langle X_H^\dagger BH^{m,n+1} V^{p,q}\rangle - iM\langle G^\dagger X_V H^{m,n} V^{p,q+1}\rangle$$
$$- iM\, p\langle X_V^\dagger BH^{m,n} V^{p-1,q}\rangle - iM\langle X_V^\dagger BH^{m,n} V^{p,q+1}\rangle + iM\langle G^\dagger X_H H^{m,n+1} V^{p,q}\rangle$$

The ground to biexciton transition does not couple the biexciton to the ground state density, but the exciton-biexciton to the ground to exciton transitions and is formally equivalent to the exciton-exciton transition. Due to this transition, the relaxation path can also interfere. Finally, the ground state density reads:

$$\partial_t \langle G^\dagger GH^{m,n} V^{p,q}\rangle \tag{6.9}$$
$$= i\left[(m-n)\omega_H^0 + (p-q)\omega_V^0 + i\kappa(m+n+p+q)\right]\langle G^\dagger GH^{m,n} V^{p,q}\rangle$$
$$- iM\, m\langle X_H^\dagger GH^{m-1,n} V^{p,q}\rangle - iM\langle X_H^\dagger GH^{m,n+1} V^{p,q}\rangle - iM\, p\langle X_V^\dagger GH^{m,n} V^{p-1,q}\rangle$$
$$- iM\langle X_V^\dagger GH^{m,n} V^{p,q+1}\rangle + iM\, n\langle G^\dagger X_H H^{m,n-1} V^{p,q}\rangle + iM\langle G^\dagger X_H H^{m+1,n} V^{p,q}\rangle$$
$$+ iM\, q\langle G^\dagger X_V H^{m,n} V^{p,q-1}\rangle + iM\langle G^\dagger X_V H^{m,n} V^{p+1,q}\rangle$$

6 Appendix

The ground state is populated by all relaxation processes as the energetic lowest electronic state. The ground state to horizontal exciton state transition reads:

$$\partial_t \langle G^\dagger X_H H^{m,n} V^{p,q} \rangle \quad (6.10)$$
$$= i \left[(m-n)\omega_H^0 + (p-q)\omega_V^0 - \omega_H + i\kappa(m+n+p+q) \right] \langle G^\dagger X_H H^{m,n} V^{p,q} \rangle$$
$$- iM\, m\, \langle X_H^\dagger X_H H^{m-1,n} V^{p,q} \rangle - iM \langle X_H^\dagger X_H H^{m,n+1} V^{p,q} \rangle$$
$$- iM\, p\, \langle X_V^\dagger X_H H^{m,n} V^{p-1,q} \rangle + iM \langle G^\dagger G H^{m,n+1} V^{p,q} \rangle + iM\, n\, \langle G^\dagger B H^{m,n-1} V^{p,q} \rangle$$
$$- iM \langle X_V^\dagger X_H H^{m,n} V^{p,q+1} \rangle + iM \langle G^\dagger B H^{m+1,n} V^{p,q} \rangle.$$

The ground to exciton transition is formally equivalent to the exciton to biexciton transition

6.3 Biexciton cascade in the weak coupling regime: Equations of motions

The equation of motion of a biexciton cascade in the weak coupling regime differ from the strong copuling regime in the number of photon operators, which are taken into account. The temporal evolution of the driving terms in Eq. (3.62) is given by

$$\partial_t \langle G^\dagger X_H c_V^\dagger c_V^\dagger c_H \rangle = \quad i\left(-\omega_H + 2\omega_0^V - \omega_0^H + i\Gamma + 3i\kappa\right) \langle G^\dagger X_H c_V^\dagger c_V^\dagger c_H \rangle \quad (6.11)$$
$$- 2iM \langle X_V^\dagger X_H c_V^\dagger c_H \rangle + iM \langle G^\dagger B c_V^\dagger c_V^\dagger \rangle$$

and

$$\partial_t \langle X_V^\dagger G\, c_V^\dagger c_H c_H \rangle = \quad i\left(\omega_V + \omega_0^V - 2\omega_0^H + i\Gamma + 3i\kappa\right) \langle X_V^\dagger G\, c_V^\dagger c_H c_H \rangle \quad (6.12)$$
$$+ 2iM \langle X_V^\dagger X_H c_V^\dagger c_H \rangle + iM \langle B^\dagger G\, c_H c_H \rangle.$$

The driving terms of the two-photon density matrix in turn couple to combined exciton- and photon coherences $X_V^\dagger X_H c_V^\dagger c_H$ and to the direct decay channel from $B^\dagger B$ to $G^\dagger G$ emitting two photons with the same polarization $G^\dagger B c_V^\dagger c_V^\dagger$. Crucial for entangling the two decay paths is the exciton-exciton transition, assisted by a photon coherence:

$$\partial_t \langle X_V^\dagger X_H c_V^\dagger c_H \rangle = \quad i\left(\omega_V - \omega_H + \omega_0^V - \omega_0^H + 2i\Gamma + 2i\kappa\right) \langle X_V^\dagger X_H c_V^\dagger c_H \rangle$$
$$- iM \langle G^\dagger X_H c_V^\dagger c_V^\dagger c_H \rangle + iM \langle B^\dagger X_H c_H \rangle$$
$$+ iM \langle X_V^\dagger G\, c_V^\dagger c_H c_H \rangle + iM \langle X_V^\dagger B c_V^\dagger \rangle. \quad (6.13)$$

In this equation, the two paths interfere. The influence in the two-particle correlation $\langle X_V^\dagger X_H c_V^\dagger c_H \rangle$ increases the degree of entanglement as this term couples back to the driving terms of ρ_{VH}, Eq. (6.11) and (6.12). Here again the resonance condition of the frequencies is essential ($\omega_V - \omega_H = \omega_0^V - \omega_0^H =$

6 Appendix

δ): A high detuning δ will diminish the contribution of Eq. (6.13) to the cascade and both paths cannot interfere. The other characteristic and important quantity in the two-electron biexciton-cascade situation are the two-photon polarizations

$$\partial_t \langle G^\dagger B c_V^\dagger c_V^\dagger \rangle = \quad i\left(-\omega_B + 2\omega_0^V + 2i\Gamma + 2i\kappa\right)\langle G^\dagger B c_V^\dagger c_V^\dagger \rangle \qquad (6.14)$$
$$+ \; iM \langle G^\dagger X_H c_V^\dagger c_V^\dagger c_H \rangle - iM \langle G^\dagger X_V c_V^\dagger c_V^\dagger c_V \rangle - 2iM \langle X_V^\dagger B c_V^\dagger \rangle$$

and

$$\partial_t \langle B^\dagger G \, c_H c_H \rangle = \quad i\left(\omega_B - 2\omega_0^H + 2i\Gamma + 2i\kappa\right)\langle B^\dagger G \, c_H c_H \rangle \qquad (6.15)$$
$$- \; iM \langle X_H^\dagger G \, c_H^\dagger c_H c_H \rangle + iM \langle X_V^\dagger G \, c_V^\dagger c_H c_H \rangle + 2iM \langle B^\dagger X_H c_H \rangle.$$

Each path in the cascade has one biexciton-to-ground state transition like $G^\dagger B c_V^\dagger c_V^\dagger$. Its dynamics couples the biexciton-to-exciton transition $X_V^\dagger B c_V^\dagger$ with both exciton-to-ground state transitions $G^\dagger X_i$. Remarkably, the origin of the entanglement is directly visible, since a quantity of a different path enters in Eq. (6.14): $\langle G^\dagger X_H c_V^\dagger c_V^\dagger c_H \rangle$. Here again, the two paths interfere. For maximum entanglement the contributions of the different paths $G^\dagger X_H$ and $G^\dagger X_V$ to the expectation values should be equally weighted. The photon-assisted biexciton-to-exciton transition enters in the two-photon polarization and drives this quantity via the biexciton decay:

$$\partial_t \langle B^\dagger X_H c_H \rangle = \quad i\left(\omega_B - \omega_H - \omega_0^H + 3i\Gamma + i\kappa\right)\langle B^\dagger X_H c_H \rangle + iM \langle B^\dagger B \rangle \qquad (6.16)$$
$$- \; iM \langle X_H^\dagger X_H c_H^\dagger c_H \rangle + iM \langle X_V^\dagger X_H c_V^\dagger c_H \rangle + iM \langle B^\dagger G \, c_H c_H \rangle,$$

$$\partial_t \langle X_V^\dagger B c_V^\dagger \rangle = \quad i\left(-\omega_B + \omega_V + \omega_0^V + 3i\Gamma + i\kappa\right)\langle X_V^\dagger B c_V^\dagger \rangle + iM \langle B^\dagger B \rangle \qquad (6.17)$$
$$- \; iM \langle G^\dagger B c_V^\dagger c_V^\dagger \rangle + iM \langle X_V^\dagger X_H c_V^\dagger c_H \rangle - iM \langle X_V^\dagger X_V c_V^\dagger c_V \rangle.$$

The occurring biexciton as well as the intermediate exciton-photon densities are driven by the biexciton-exciton transition $\langle X_i^\dagger B c_i^\dagger \rangle$:

$$\partial_t \langle X_H^\dagger X_H c_H^\dagger c_H \rangle = \; - \; (2\Gamma + 2\kappa) \langle X_H^\dagger X_H c_H^\dagger c_H \rangle \qquad (6.18)$$
$$- \; 2 \, \mathrm{Im} \left(M \langle X_H^\dagger B c_H^\dagger \rangle + M \langle X_H^\dagger G \, c_H^\dagger c_H c_H \rangle \right),$$
$$\partial_t \langle X_V^\dagger X_V c_V^\dagger c_V \rangle = \; - \; (2\Gamma + 2\kappa) \langle X_V^\dagger X_V c_V^\dagger c_V \rangle \qquad (6.19)$$
$$+ \; 2 \, \mathrm{Im} \left(M \langle X_V^\dagger B c_V^\dagger \rangle - M \langle X_V^\dagger G \, c_V^\dagger c_V c_V \rangle \right).$$

In the visualization of the complex interplay, it is a bottom-to-top trail followed through the cascade, starting with the concurrence determining ρ_{VH}. The biexciton $\langle B^\dagger B\rangle$ as the top element of the scheme decays via the H or the V intermediate exciton-to-ground-state path

$$\partial_t\langle B^\dagger B\rangle = -4\Gamma\langle B^\dagger B\rangle + 2\,\text{Im}\left(M\langle X_H^\dagger Bc_H^\dagger\rangle - M\langle X_V^\dagger Bc_V^\dagger\rangle\right). \tag{6.20}$$

To complete the set of equation, two higher-order photon-assisted exciton-to-ground state transitions of the direct and thus not entangled path are necessary:

$$\partial_t\langle G^\dagger X_H c_H^\dagger c_H^\dagger c_H\rangle = i\left(-\omega_H + \omega_0^H + i\Gamma + 3i\kappa\right)\langle G^\dagger X_H c_H^\dagger c_H^\dagger c_H\rangle \tag{6.21}$$
$$- 2iM\langle X_H^\dagger X_H c_H^\dagger c_H\rangle + iM\langle G^\dagger Bc_H^\dagger c_H^\dagger\rangle,$$
$$\partial_t\langle G^\dagger X_V c_V^\dagger c_V^\dagger c_V\rangle = i\left(-\omega_V + \omega_0^V + i\Gamma + 3i\kappa\right)\langle G^\dagger X_V c_V^\dagger c_V^\dagger c_V\rangle \tag{6.22}$$
$$- 2iM\langle X_V^\dagger X_V c_V^\dagger c_V\rangle - iM\langle G^\dagger Bc_V^\dagger c_V^\dagger\rangle.$$

With these polarization Eq. (6.21-6.22), the diagonal elements $i = H, V$ of the density matrix of the polarization subspace are given, too:

$$\partial_t\langle c_i^\dagger c_i^\dagger c_i c_i\rangle = -4\kappa\langle c_i^\dagger c_i^\dagger c_i c_i\rangle - 4\,\text{Im}\left(M\langle G^\dagger X_i^\dagger c_i^\dagger c_i^\dagger c_i\rangle\right). \tag{6.23}$$

6.4 Derivation of the Hartree-Fock factorization

To close the set of differential equations, one needs to factorize the expectation values at a given level, e.g. on a Hartree-Fock level: $\langle a_i^\dagger a_j^\dagger a_k a_l\rangle \approx \langle a_i^\dagger a_l\rangle\langle a_j^\dagger a_k\rangle - \langle a_i^\dagger a_k\rangle\langle a_j^\dagger a_l\rangle$. This factorization rule is derived via the approximation, that the electron-electron correlations can be expressed as a general canonical statistical operator (GCSO), including only single particle contributions.[Fri96; FS90] That is: the single particle dynamics is described in a mean field, induced by the other particles. The GCSO reads:

$$\rho \approx \rho_{HF} = \frac{1}{Z}e^{-\sum_{ij}\lambda_{ij}a_i^\dagger a_j} \quad , \quad Z = \text{tr}(e^{-\sum_{ij}\lambda_{ij}a_i^\dagger a_j}), \tag{6.24}$$

with the partition function as a normalization to secure $\text{tr}(\rho) = 1$ and $a_i^{(\dagger)}$ fermionic operators with $[a_i^\dagger, a_j]_+ = \delta_{ij}$ and $a_i^\dagger a_j^\dagger = -a_j^\dagger a_i^\dagger$. Now, one introduces a unitary matrix ϕ to diagonalize, so that $\lambda^D = \phi\lambda\phi^*$, which exist as long as $H_{HF} = \sum_{ij}\lambda_{ij}a_i^\dagger a_j$ consists of observables, i.e. H_{HF} is hermitian. The matrix element reads:

$$\lambda_{ij} = \sum_{mn}\phi_{in}^*\lambda_{nm}^D\phi_{mj} = \sum_n \phi_{in}^*\lambda_{nn}^D\phi_{nj}. \tag{6.25}$$

6 APPENDIX

The GCSO can now be written in this new basis and is diagonal:

$$\rho_{HF} = \frac{1}{Z} e^{-\sum_n \lambda_{nn}^D d_n^\dagger d_n}, \qquad (6.26)$$

with the new operators

$$d_n^{(\dagger)} = \sum_i \phi_{in}^{(*)} a_i^{(\dagger)}, \qquad (6.27)$$

inheriting the fermionic character. The partition function can be calculated, given a complete set of eigenfunction of the new operators:

$$|\Psi_N\rangle := |n_1, n_2, n_3, ..., n_N\rangle = d_1^\dagger d_2^\dagger \cdots d_N^\dagger |0\rangle = -|n_2, n_1, n_3, ..., n_N\rangle. \qquad (6.28)$$

With this definition, one can calculate with P_i the number of necessary anti-commutations to bring the state i to the front:

$$d_i^\dagger d_i |\Psi_N\rangle = d_i^\dagger d_i (-1)^{P_i} |n_i, n_1, ..., n_{i-1}, n_{i+1}, ..., n_N\rangle = n_i \left[(-1)^{P_i}\right]^2 |\Psi_N\rangle = n_i |\Psi_N\rangle. \qquad (6.29)$$

The trace is the summation of every state, which can be occupied $n_i = 1$ or be unoccupied $n_i = 0$, taken into account Pauli's principle. The trace of the diagonalized operator reads:

$$\begin{aligned} Z &= \sum_{\{n_i\}} \langle \Psi_N | e^{-\sum_n \lambda_{nn}^D d_n^\dagger d_n} |\Psi_N\rangle = \sum_{n_1=0}^{1} \sum_{n_2=0}^{1} \cdots \sum_{n_N=0}^{1} \langle \Psi_N | e^{-\sum_n \lambda_{nn}^D d_n^\dagger d_n} |\Psi_N\rangle \\ &= \sum_{n_1=0}^{1} \sum_{n_2=0}^{1} \cdots \sum_{n_N=0}^{1} \langle \Psi_N | e^{-\sum_{n=1}^{N-1} \lambda_{nn}^D d_n^\dagger d_n} |\Psi_N\rangle e^{-\lambda_{NN}^D n_N} \\ &= \sum_{n_1=0}^{1} \sum_{n_2=0}^{1} \cdots \sum_{n_N=0}^{1} e^{-\lambda_{11}^D n_1} e^{-\lambda_{22}^D n_2} \cdots e^{-\lambda_{NN}^D n_N} \\ &= (1 + e^{-\lambda_{11}^D})(1 + e^{-\lambda_{22}^D}) \cdots (1 + e^{-\lambda_{NN}^D}) = \Pi_k (1 + e^{-\lambda_{kk}^D}). \end{aligned} \qquad (6.30)$$

Now, the GCSO is explicitly given and one can calculate on a Hartree-Fock level the expectation value of the electron-electron correlations. The approximation is done in the choice of the GCSO, in which the observables of interest are included, here single particle contribution.

$$\langle a_i^\dagger a_j^\dagger a_k a_l \rangle = \text{tr}(a_i^\dagger a_j^\dagger a_k a_l \rho) \approx \sum_{abcd} \phi_{ai} \phi_{bj} \phi_{kc}^* \phi_{ld}^* \text{tr}(\rho_{HF} d_a^\dagger d_b^\dagger d_c d_d). \qquad (6.31)$$

The trace must now be taken into account, for that, one investigates:

$$d_a^\dagger d_b^\dagger d_c d_d |n_d, n_c, ..., n_N\rangle = \sqrt{n_c n_d} \langle n_c, n_d, ... n_N | d_a^\dagger d_b^\dagger |n_d - 1, n_c - 1, ..., n_N\rangle \qquad (6.32)$$

Summing over b results only in two contributions, since the states are orthogonal:

$$\begin{aligned} d_a^\dagger d_b^\dagger d_c d_d |\Psi_N\rangle &= \sqrt{n_d} n_c \, d_a^\dagger |n_d - 1, n_c, ..., n_N\rangle \delta_{b,c} \\ &\quad + n_d \sqrt{n_c} \, d_a^\dagger |n_d, n_c - 1, ..., n_N\rangle \delta_{b,d} \\ &= n_d n_c |\Psi_N\rangle (\delta_{a,d}\delta_{b,c} - \delta_{a,c}\delta_{b,d}). \end{aligned} \qquad (6.33)$$

Now, the expectation value can be calculated:

$$\begin{aligned} \langle a_i^\dagger a_j^\dagger a_k a_l \rangle &\approx \sum_{\{n_i\}} \sum_{abcd} \phi_{ai} \phi_{bj} \phi_{kc}^* \phi_{ld}^* \langle \Psi_N | \rho_{HF} \, d_a^\dagger d_b^\dagger d_c d_d |\Psi_N\rangle \qquad (6.34)\\ &= \sum_{cd} \sum_{\{n_i\}} n_c \, n_d \, \langle \Psi_N | \rho_{HF} |\Psi_N\rangle \left(\phi_{di}\phi_{cj}\phi_{kc}^*\phi_{ld}^* - \phi_{ci}\phi_{dj}\phi_{kc}^*\phi_{ld}^* \right) \\ &= \sum_{cd} \sum_{n_c, n_d = 0}^{1} \left[\frac{n_c \, e^{-\lambda_{cc}^D n_c}}{(1 + e^{-\lambda_{cc}^D})} \right] \left[\frac{n_d \, e^{-\lambda_{dd}^D n_d}}{(1 + e^{-\lambda_{dd}^D})} \right] \left(\phi_{di}\phi_{cj}\phi_{kc}^*\phi_{ld}^* - \phi_{ci}\phi_{dj}\phi_{kc}^*\phi_{ld}^* \right) \\ &= \sum_{cd} \left[\frac{\phi_{di} e^{-\lambda_{cc}^D} \phi_{ld}^*}{(1 + e^{-\lambda_{cc}^D})} \right] \left[\frac{\phi_{cj} e^{-\lambda_{dd}^D} \phi_{kc}^*}{(1 + e^{-\lambda_{dd}^D})} \right] - \left[\frac{\phi_{dj} e^{-\lambda_{cc}^D} \phi_{ld}^*}{(1 + e^{-\lambda_{cc}^D})} \right] \left[\frac{\phi_{ci} e^{-\lambda_{dd}^D} \phi_{kc}^*}{(1 + e^{-\lambda_{dd}^D})} \right] \\ &= \langle a_i^\dagger a_l \rangle \langle a_j^\dagger a_k \rangle - \langle a_i^\dagger a_k \rangle \langle a_j^\dagger a_l \rangle. \end{aligned}$$

After choosing the Hartree-Fock GCSO, the calculation leads to automatically to the factorization.

6.5 Photon probability picture

In Eq. (4.14) a new form of the phase-filling factor is derived. The spontaneous emission and induced emission is included in the quantity $\langle a_c^\dagger a_c | n \rangle \langle n |\rangle$, whereas the induced absorption is included in $\langle a_v^\dagger a_v | n + 1 \rangle \langle n + 1 |\rangle$. To see that, one can transform Eq. (4.14) back into the photon operator picture. E.g., one sums over n after multiplying by $\sqrt{n+1}$, keeping in mind the annihilation and creation operator relations:

$$\begin{aligned} c^\dagger |n\rangle &= \sqrt{n+1} \; |n+1\rangle \quad \equiv \quad \langle n | c = \langle n+1 | \sqrt{n+1}, & (6.35) \\ c \, |n\rangle &= \sqrt{n} \; |n-1\rangle \quad \equiv \quad \langle n | c^\dagger = \langle n-1 | \sqrt{n}. & (6.36) \end{aligned}$$

With that, one transforms the equation of the photon-assisted polarization part by part back:

$$\sum_n \sqrt{n+1} \langle a_v^\dagger a_c | n+1 \rangle \langle n |\rangle = \sum_n \langle a_v^\dagger a_c c^\dagger | n \rangle \langle n |\rangle = \langle a_v^\dagger a_c c^\dagger \left(\sum_n | n \rangle \langle n | \right) \rangle = \langle a_v^\dagger a_c c^\dagger \rangle.$$

6 APPENDIX

The inhomogeneity in Eq. (4.14), are already multiplied by $\sqrt{n+1}$, so we have:

$$\sum_n (n+1)\langle a_c^\dagger a_c |n\rangle\langle n|\rangle = \sum_n \langle a_c^\dagger a_c |n\rangle\langle n|\rangle + n\langle a_c^\dagger a_c |n\rangle\langle n|\rangle \tag{6.37}$$

$$= \langle a_c^\dagger a_c\rangle + \sum_n \langle a_c^\dagger a_c c^\dagger c|n\rangle\langle n|\rangle = \langle a_c^\dagger a_c\rangle + \langle a_c^\dagger a_c c^\dagger c\rangle$$

and

$$\sum_n (n+1)\langle a_v^\dagger a_v |n+1\rangle\langle n+1|\rangle = \sum_n \langle a_v^\dagger a_v c^\dagger |n\rangle\langle n|c\rangle = \langle a_v^\dagger a_v c^\dagger c\rangle. \tag{6.38}$$

We come up with the photon-assisted polarization, in the second order of the electron-light coupling in its known form:

$$\partial_t \langle a_v^\dagger a_c c^\dagger\rangle = -i(\omega_{cv} - \omega_0)\langle a_v^\dagger a_c c^\dagger\rangle - i M \left(\langle a_c^\dagger a_c\rangle + \langle a_c^\dagger a_c c^\dagger c\rangle - \langle a_v^\dagger a_v c^\dagger c\rangle\right), \tag{6.39}$$

and now the spontaneous emission term, proportional to the excited state density, and induced absorption and emission in the phase-filling factor. Remarkable, in dependence on the factor one multiplies the equation, one automatically derives different order in the electron-photon coupling element. As an example, we investigate the photon probability $\langle |n\rangle\langle n|\rangle$:

$$\sum_n n \langle |n\rangle\langle n|\rangle = \langle c^\dagger c\rangle \tag{6.40}$$

$$\sum_n n(n-1) \langle |n\rangle\langle n|\rangle = \langle c^\dagger c^\dagger cc\rangle. \tag{6.41}$$

This works for every expectation value. Thus, one needs for the whole dynamics in arbitrary orders only four equations of motions.

6.6 Jaynes-Cummings solution in the photon probability picture

In the one-electron assumption without losses and other interactions, the exact solution is given by the JCM. In case of the vacuum Rabi oscillation, i.e. zero photons in the cavity and a initially excited two-level system, the photon density reads: $\langle c^\dagger c\rangle(t) = \sin^2(M\,t)$.[JC63] We start by showing that the PPCE reproduces this solution. In the one-electron case, the photon-assisted polarization reads for $\omega_{cv} - \omega_0 = 0$:

$$\partial_t \langle |n+1\rangle\langle n|a_v^\dagger a_c\rangle = +i M \sqrt{n+1}(p_{n+1} - f_{n+1}^h) - i M \sqrt{n+1}f_n^e. \tag{6.42}$$

6 Appendix

Here, we use the electron-hole picture. Note, the n-quantity considers spontaneous emission, since the dynamics couple to a lower photon probability in comparison to the photon-assisted polarization and in contrast to the $n + 1$-quantities. The equation of motion for the photon-assisted electron and hole densities couple again to the photon-assisted polarization:

$$\partial_t f_n^{e/h} = 2\sqrt{n+1}\,\text{Im}(M \langle |n + 1\rangle\langle n|a_v^\dagger a_c\rangle). \quad (6.43)$$

For fixed initial conditions, without environment coupling, the set of equations [Eq. (6.42) and 6.43] can be solved analytically, e.g. in the case of vacuum Rabi oscillation. The initial conditions are: $f_0^e = f_0^h = 1, p_0 = 1, p_1 = 0$. This solution is obtained by differentiating Eq. (6.43) for $n = 0$ and (4.13) for $n = 1$ with respect to time:

$$\partial_t^2 f_0^{e/h} = 2\text{Im}(M \,\partial_t \langle |1\rangle\langle 0|a_v^\dagger a_c\rangle) = -\partial_t^2 p_1. \quad (6.44)$$

This is valid, as long, as higher order photon assisted polarization are not driven, i.e. $\langle a_v^\dagger a_c|n + 1\rangle\langle n|\rangle$ for $n > 0$. Investing the next higher photon assisted polarization $n = 1$:

$$\partial_t \langle |2\rangle\langle 1|a_v^\dagger a_c\rangle = +i\,M\,\sqrt{2}(p_2 - f_2^h) - i\,M\,\sqrt{2}f_1^e, \quad (6.45)$$

which is initially zero. To start the dynamics, one of the inhomogeneties must be inequal to zero at least at one time. f_1^e is initially zero and couples back to $\langle |2\rangle\langle 1|a_v^\dagger a_c\rangle$ and cannot start the dynamics; the same for f_2^h, which couples to a even higher photon assisted polarization and is initially zero as well. At last, p_2 needs to be investigated. Since there is initially zero photons in the cavity, and there is only one excitation in the system, it is physically impossible to come up with a probability for two photons, in the rotating wave assumption and considering a band gap of the quantum dot of at least order of magnitude eV. Time and energy uncertainty forbids an existence of an photon longer than a tenth of a femtosecond. But it is also mathematically clear, p_2 couples to the photon-assisted polarization of $\langle |3\rangle\langle 2|a_v^\dagger a_c\rangle$, which gives clearly no contribution, and couples back to $\langle |2\rangle\langle 1|a_v^\dagger a_c\rangle$. As long as initially p_2 is zero, higher photon assisted polarization can be neglected and Eq. (6.44) is valid and the photon assisted polarization of $n = 0$ reads:

$$\partial_t \langle |1\rangle\langle 0|a_v^\dagger a_c\rangle = +i\,M\,(p_1 - f_0^e), \quad (6.46)$$

since f_1^h is not driven and initially zero, it couples to Eq. (6.45). Obviously, it is now convenient, to investigate the dynamics of the difference between the electron density of n_0 and the photon probability of $n = 1$, this

$$\partial_t^2 (f_0^e - p_1) = 2\text{Im}(M\,\partial_t\langle |1\rangle\langle 0|a_v^\dagger a_c\rangle) - \partial_t^2 p_1 \quad (6.47)$$
$$= 4\text{Im}(M\,\partial_t\langle |1\rangle\langle 0|a_v^\dagger a_c\rangle)$$

6 Appendix

After inserting the photon-assisted polarization for $n = 0$ from Eq. (6.42), the set of equations of motions is closed, since all other quantities in higher photon manifolds are not driven with given initial conditions:

$$\partial_t^2 (f_0^e - p_1) = -(2M)^2 (f_0^e - p_1). \qquad (6.48)$$

The solution reads $f_0^e(t) - p_1(t) = \cos(2M\,t) = \cos^2(M\,t) - \sin^2(M\,t)$. Since $f_0^e(t=0) = 1$, one identifies $f_0^e(t) = \cos^2(M\,t)$ and $p_1(t) = \sin^2(M\,t)$. With the initial conditions, the photon density depends only on p_1 and p_0 and reads:

$$\langle c^\dagger c \rangle(t) = = 0 \cdot p_0 + 1 \cdot p_1 = p_1(t) = \sin^2(M\,t), \qquad (6.49)$$

the JCM solution is reproduced. The oscillation frequency depends on the number of photons in the cavity and the off-diagonal light coupling M^{vc}, forming together the Rabi frequency [JC63]. Thus, the JCM is naturally contained in the PPCE-approach.[RCSK09a] Another benchmark is to include the photon-statistics via initial conditions like in the JCM, which is done in the next subsection.

6.7 Modified photon-assisted polarization (electron picture)

The photon-assisted polarization reads in the electron picture:

$$\partial_t \langle a_v^\dagger a_c | n+1 \rangle \langle n | \rangle \qquad (6.50)$$
$$= -i(\omega_{cv} - \omega_0)\langle a_v^\dagger a_c | n+1 \rangle \langle n | \rangle$$
$$+ i M \sqrt{n+1} \langle a_v^\dagger a_v | n+1 \rangle \langle n+1 | \rangle \left(1 - \frac{1}{p_{n+1}} \langle a_c^\dagger a_c | n+1 \rangle \langle n+1 | \rangle\right)$$
$$- i M \sqrt{n+1} \langle a_c^\dagger a_c | n \rangle \langle n | \rangle \left(1 - \frac{1}{p_n} \langle a_v^\dagger a_v | n \rangle \langle n | \rangle\right).$$

6.8 Cavity loss in the photon probability picture

The coupling of the cavity modes to the external modes is decribed by an additional part in the Hamiltonoperator,[VWW01] which reads together with the homogenous part of the external photons:

$$H_{pt-pt} = \hbar \sum_q \omega_q d_q^\dagger d_q + \hbar \sum_q G_q (c^\dagger + c)(d_q^\dagger + d_q), \qquad (6.51)$$

The probability of the tunneling process from the cavity into the environment is determined by the strength of the coupling element G_q of the mode q. For the cavity loss external photon modes described by $d_q, d_q^\dagger$. Only contributions within the rotating-wave approximation are considered. The dynamics,

6 APPENDIX

calculated only for the photon-photon interaction reads for an arbitrary QD operator A assisted by $|n\rangle\langle m|$ cavity photons:

$$\partial_t \langle A|n\rangle\langle m|\rangle|_{pt-pt} = i(n-m)\omega_0 \langle A|n\rangle\langle m|\rangle \qquad (6.52)$$
$$+ i\sum_q G_q \left(\sqrt{n+1} \langle A|n+1\rangle\langle m|d_q\rangle + \sqrt{n} \langle A|n-1\rangle\langle m|d_q^\dagger\rangle \right)$$
$$+ i\sum_q G_q \left(\sqrt{m+1} \langle A|n\rangle\langle m+1|d_q^\dagger\rangle + \sqrt{m} \langle A|n\rangle\langle m-1|d_q\rangle \right).$$

Now, a rotating frame is used:

$$\langle A|n\rangle\langle m|\rangle = \langle A|n\rangle\langle m|\rangle_R \, e^{i(m-n)\omega_0 t} \qquad (6.53)$$
$$\langle A|n\rangle\langle m|d_q\rangle = \langle A|n\rangle\langle m|d_q\rangle_R \, e^{i((m-n)\omega_0 - \omega_q)t} \qquad (6.54)$$

Now, the free part is changed into a phase-factor:

$$\partial_t \langle A|n\rangle\langle m|\rangle_R|_{pt-pt} = i\sum_q G_q \sqrt{n+1} \langle A|n+1\rangle\langle m|d_q\rangle_R \, e^{i(\omega_0 - \omega_q)t} \qquad (6.55)$$
$$+ i\sum_q G_q \sqrt{n} \langle A|n-1\rangle\langle m|d_q^\dagger\rangle_R \, e^{-i(\omega_0 - \omega_q)t}$$
$$+ i\sum_q G_q \sqrt{m+1} \langle A|n\rangle\langle m+1|d_q^\dagger\rangle_R \, e^{-i(\omega_0 - \omega_q)t}$$
$$+ i\sum_q G_q \sqrt{m} \langle A|n\rangle\langle m-1|d_q\rangle_R \, e^{i(\omega_0 - \omega_q)t}.$$

Now, the dissipative photon mode assisted quantities must be calculated, to derive the set of equation up to the second order in the tunneling matrix couling element. They read:

$$\partial_t \langle A|n\rangle\langle m|d_q\rangle_R|_{pt-pt} = i\sum_{q'} G_{q'} \sqrt{n} \langle A|n-1\rangle\langle m|d_{q'}^\dagger d_q\rangle_R \, e^{-i(\omega_0 - \omega_{q'})t} \qquad (6.56)$$
$$- i\sum_{q'} G_{q'} \sqrt{m+1} \langle A|n\rangle\langle m+1|\rangle_R \, e^{-i(\omega_0 - \omega_{q'})t} \delta_{q,q'}$$
$$- i\sum_q G_q \sqrt{m+1} \langle A|n\rangle\langle m+1|d_{q'}^\dagger d_q\rangle_R \, e^{-i(\omega_0 - \omega_{q'})t}$$
$$\approx -i G_q \sqrt{m+1} \langle A|n\rangle\langle m+1|\rangle_R \, e^{-i(\omega_0 - \omega_q)t}.$$

Contributions proportional to $\langle d_q^{(\dagger)} d_{q'}^{(\dagger)} \rangle$ are neglected within the bath assumption, since only dissipative are taken into account. The bath photon do not couple back into the cavity. The mean photon number is too low in the bath to consider a tunnel process in the reversed direction. The phase-relation is the same for the inhomogeneities, since this contributions reflects one part of the two part density matrix process for the cavity loss, the other one reads:

$$\partial_t \langle A|n\rangle\langle m|d_q^\dagger\rangle_R|_{pt-pt} \approx i G_q \sqrt{n+1} \langle A|n+1\rangle\langle m|\rangle_R \, e^{i(\omega_0 - \omega_q)t}. \qquad (6.57)$$

6 APPENDIX

Now, the equation is solved by integrating formally and applying the Markovian approximation, assuming memory effects are negligible, which is a good approximation in system-bath processes, with coupling strength suggesting interaction time higher than ps-scale. The tunnel coupling strength is weak enough and the solution reads:

$$\langle A|n\rangle\langle m|d_q^\dagger\rangle_R(t)|_{pt-pt} \approx iG_q \sqrt{n+1}\langle A|n+1\rangle\langle m|\rangle_R \int_0^\infty d\tau\, e^{i(\omega_0-\omega_q)(t-\tau)} \quad (6.58)$$

$$= iG_q \sqrt{n+1}\langle A|n+1\rangle\langle m|\rangle_R\, e^{i(\omega_0-\omega_q)t} \int_0^\infty d\tau\, e^{-i(\omega_0-\omega_q)\tau}.$$

If the lower limit is not extended from 0 to $-\infty$, the integral is the Heitler-Zeta function $\zeta(\omega_0-\omega_q)$, consisting of the delta-function and of the Cauchy-principal value integral. It must be checked, to what extent the Cauchy-principal vaule integral contributes to the dynamics, normally a energy renormalization occurs, which can be included into the definition of ω_0. In Eq. (6.56), several combinations of dissipative photon mode assisted quantities need to be calculated like in Eq. (4.60). They read:

$$\langle A|n+1\rangle\langle m|d_q\rangle_R = -i\, G_q \sqrt{m+1}\langle A|n+1\rangle\langle m+1|\rangle_R\, e^{-i(\omega_0-\omega_q)t}\zeta(\omega_0-\omega_q) \quad (6.59)$$

$$\langle A|n-1\rangle\langle m|d_q^\dagger\rangle_R = i\, G_q \sqrt{n}\langle A|n\rangle\langle m|\rangle_R\, e^{i(\omega_0-\omega_q)t}\zeta(\omega_0-\omega_q)$$

$$\langle A|n\rangle\langle m-1|d_q\rangle_R = -i\, G_q \sqrt{m}\langle A|n\rangle\langle m|\rangle_R\, e^{-i(\omega_0-\omega_q)t}\zeta(\omega_0-\omega_q)$$

$$\langle A|n\rangle\langle m+1|d_q^\dagger\rangle_R = i\, G_q \sqrt{n}\langle A|n+1\rangle\langle m+1|\rangle_R\, e^{i(\omega_0-\omega_q)t}\zeta(\omega_0-\omega_q)$$

These solutions can be plug into the Eq. (6.56) to derive the cavity-loss for the general quantity $\langle A|n\rangle\langle m|\rangle$, by defining:

$$\kappa := \sum_q |G_q|^2\, \zeta(\omega_0-\omega_q), \quad (6.60)$$

the PPCE formulation of the Markovian cavity loss, including energy renormalization is derived and can be expressed as:

$$\partial_t\langle|n\rangle\langle m|A\rangle_R|_{ph-ph} = -(m+n)\kappa\langle|n\rangle\langle m|A\rangle$$
$$+2\sqrt{m+1}\sqrt{n+1}\kappa\langle|n+1\rangle\langle m+1|A\rangle. \quad (6.61)$$

In the PPCE formalism, the dynamics of the photon loss describes the loss and gain of photons of the different photon probabilities self-consistently. The dynamics of the photon probabilities are easier in the electron-photon interaction, but more complicated for the photon-loss, since inscattering from higher order photon probabilities need to be considered. Therefore, the cavity loss appears more complicated than in the picture of the $c^\dagger, c$-operators. To give an example, one chooses in Eq. (4.62) the photon-assisted transition. The equation reads, only for the photon-photon interaction:

$$\partial_t\langle a_v^\dagger a_c|n+1\rangle\langle n|\rangle_R = -(2n+1)\kappa\langle a_v^\dagger a_c|n+1\rangle\langle n|\rangle_R + 2\sqrt{n+1}\sqrt{n+2}\kappa\langle a_v^\dagger a_c|n+2\rangle\langle n+1|\rangle_R.$$

6 Appendix

To transform Eq. (6.62) into the operator formalism, one multiplies, e.g. with $\sqrt{n+1}$, and sums over n:

$$\sum_n \sqrt{n+1}\,\partial_t \langle a_v^\dagger a_c |n+1\rangle\langle n|\rangle_R = \sum_n \partial_t \langle a_v^\dagger a_c |n\rangle\langle n|\rangle_R = \partial_t \langle a_v^\dagger a_c c^\dagger \rangle_R$$

$$-\sum_n \sqrt{n+1}(2n+1)\kappa \langle a_v^\dagger a_c |n+1\rangle\langle n|\rangle_R = -\kappa \langle a_v^\dagger a_c c^\dagger \rangle_R - 2\kappa \langle a_v^\dagger a_c c^\dagger c^\dagger c\rangle$$

$$\sum_n 2\sqrt{n+1}\sqrt{n+2}\kappa \langle a_v^\dagger a_c |n+2\rangle\langle n+1|\rangle_R = 2\kappa \langle a_v^\dagger a_c c^\dagger c^\dagger c\rangle_R.$$

Only with the inscattering of photons from higher photon-probabilities in case of losses, the usual cavity loss can be derived, since Eq. (6.62) reads now:

$$\partial_t \langle a_v^\dagger a_c c^\dagger \rangle_R = -\kappa \langle a_v^\dagger a_c c^\dagger \rangle_R. \tag{6.62}$$

The last equation has been derived to clarify the processes, which do not appear in the operator-formalism, that the photon-probability approach opens up further insight into the combined electron-photon dynamics, even in the case of the Markovian-derived cavity loss. The few-electron approach, i.e. the need to factorize in the electronic system, does not enter in this derivation of the cavity loss, since the cavity loss results solely from the photon-photon interaction.

6 Appendix

Danksagung

Ich danke Prof. Andreas Knorr für sein Vertrauen, seine Unterstützung und Ermutigungen und dafür, meinen verschiedenen Projekten so viel Aufmerksamkeit gewidmet zu haben, dass sie in dieser kurzen Zeit gelingen konnten. Ich danke in diesem Sinne ebenfalls Dr. Marten Richter, der mir mit Rat und Tat bei allen Problemen zur Seite gestanden hat und ohne den diese vorliegende Arbeit ebenfalls gar nicht möglich gewesen wäre, sowie der gesamten Arbeitsgruppe Knorr für die freundliche, interessante und inspirierende Arbeitsatmosphäre.

Mein Dank geht auch an Prof. Torsten Meier für die Übernahme des Zweitgutachters sowie an Prof. Michael Kneissl, dass er den Prüfungsvorsitz übernommen hat.

Ich danke Prof. Weng W. Chow für die zahlreichen fruchtbaren Diskussion hier in Berlin und in Albuquerque, wo er zusammen mit Prof. Andreas Knorr Julia Kabuss und mir einen einmonatigen Forschungsaufenthalt ermöglicht hat.

Für das Korrekturlesen und die unermüdliche Bereitschaft, wissenschaftliche Diskussionen zu führen, seien Marten Richter, Carsten Weber, Matthias-Rene Dachner, Mario Schoth, Yumian Su und vor allem Julia Kabuss gedankt. Unzählige Male habt ihr mir geholfen und euch für meine Fragen Zeit genommen. In diesem Sinne danke ich auch Jeong-Eun Kim und Ermin Malic für ihre Freundlichkeit und Verlässlichkeit.

Matthias-Rene Dachner danke ich für die temperaturabhängigen T_1-Zeiten für das Verschränkungsprojekt. An Mario Schoth geht mein Dank dafür, dass er das NLPF-Projekt so erfolgreich übernommen hat.

Meinen Eltern möchte ich herzlich danken, dass sie mir dieses wunderbare Studium mit seinen Irrungen und Wirrungen ermöglicht haben und nicht ein einziges Mal an meiner wissenschaftlichen Ernsthaftigkeit und meinem Wissensdurst zweifelten. Und vor allen anderen danke ich meiner lieben Julia, dafür, dass sie so ist, wie sie ist, für dieses gemeinsame Leben, das ich mir gar nicht schöner vorstellen kann.

Bibliography

AFK05 AHN, Kwang J. ; FÖRSTNER, Jens ; KNORR, Andreas: Resonance fluorescence of semiconductor quantum dots: Signatures of the electron-phonon interaction. In: *Phys. Rev. B* 71 (2005), Nr. 15, S. 153309

AHK99 AXT, V. M. ; HERBST, M. ; KUHN, T.: Coherent control of phonon quantum beats. In: *Superlattices and Microstructures* 26 (1999), Nr. 2, S. 117 – 128

AKVP05 AXT, V. M. ; KUHN, T. ; VAGOV, A. ; PEETERS, F. M.: Phonon-induced pure dephasing in exciton-biexciton quantum dot systems driven by ultrafast laser pulse sequences. In: *Phys. Rev. B* 72 (2005), Nr. 12, S. 125309

AVB+09 ASSMANN, M. ; VEIT, F. ; BAYER, M. ; POEL, M. van d. ; HVAM, J. M.: Higher-Order Photon Bunching in a Semiconductor Microcavity. In: *Science* 325 (2009), Nr. 5938, S. 297–300

BAS+06 BENNETT, A. J. ; ATKINSON, P. ; SEE, P. ; WARD, M. B. ; STEVENSON, R. M. ; YUAN, Z. L. ; UNITT, D. C. ; ELLIS, D. J. P. ; COOPER, K. ; RITCHIE, D. A. ; SHIELDS, A. J.: Single-photon-emitting diodes: a review. In: *phys. status solidi (b)* 243 (2006), Nr. 14, S. 3730–3740

BBG+02 BEVERATOS, Alexios ; BROURI, Rosa ; GACOIN, Thierry ; VILLING, André ; POIZAT, Jean-Philippe ; GRANGIER, Philippe: Single Photon Quantum Cryptography. In: *Phys. Rev. Lett.* 89 (2002), Oct, Nr. 18, S. 187901

BDSW96 BENNETT, Charles H. ; DIVINCENZO, David P. ; SMOLIN, John A. ; WOOTTERS, William K.: Mixed-state entanglement and quantum error correction. In: *Phys. Rev. A* 54 (1996), Nov, Nr. 5, S. 3824–3851

Bel64 BELL, J.S.: On the einstein-podolsky-rosen paradox. In: *Physics* 1 (1964), Nr. 3, S. 195–200

BGWJ06 BAER, N. ; GIES, C. ; WIERSIG, J. ; JAHNKE, F.: Luminescence of a semiconductor quantum dot system. In: *Eur. Phys. J. B* 50 (2006), Nr. 3, S. 411–418

BH84 BEACH, R. ; HARTMANN, S.R.: Incoherent photon echoes. In: *Phys. Rev. Lett.* 53 (1984), Nr. 7, S. 663 – 666

BLS+02 BORRI, P. ; LANGBEIN, W. ; SCHNEIDER, S. ; WOGGON, U. ; SELLIN, R. L. ; OUYANG, D. ; BIMBERG, D.: Relaxation and Dephasing of Multiexcitons in Semiconductor Quantum Dots. In: *Phys. Rev. Lett.* 89 (2002), Oct, Nr. 18, S. 187401

BO06 BASANO, Lorenzo ; OTTONELLO, Pasquale: Experiment in lensless ghost imaging with thermal light. In: *Applied Physics Letters* 89 (2006), aug., Nr. 9, S. 091109

BPM⁺97 BOUWMEESTER, D. ; PAN, J.W. ; MATTLE, K. ; EIBL, M. ; WEINFURTER, H. ; ZEILINGER, A.: Experimental quantum teleportation. In: *Nature* 390 (1997), S. 575–579

BSKM⁺96 BRUNE, M. ; SCHMIDT-KAHLER, F. ; MAALI, A. ; DREYER, J. ; HAGLEY, E. ; RAIMOND, J.M. ; HAROCHE, S.: Quantum Rabi Oscillation: A Direct Test of Field Quantization in a Cavity. In: *Phys. Rev. Lett.* 76 (1996), S. 1800

BSL⁺09 BIMBERG, D. ; STOCK, E. ; LOCHMANN, A. ; SCHLIWA, A. ; TOFFLINGER, J.A. ; UNRAU, W. ; MUNNIX, M. ; RODT, S. ; HAISLER, V.A. ; TOROPOV, A.I. ; BAKAROV, A. ; KALAGIN, A.K.: Quantum Dots for Single- and Entangled-Photon Emitters. In: *IEEE Photonics J.* 1 (2009), June, Nr. 1, S. 58–68

BSPY00 BENSON, Oliver ; SANTORI, Charles ; PELTON, Matthew ; YAMAMOTO, Yoshihisa: Regulated and Entangled Photons from a Single Quantum Dot. In: *Phys. Rev. Lett.* 84 (2000), Mar, Nr. 11, S. 2513–2516

BUA⁺05 BENNETT, A. ; UNITT, D. ; ATKINSON, P. ; RITCHIE, D. ; SHIELDS, A.: High performance single photon sources from photolithographically defined pillar microcavities. In: *Opt. Express* 13 (2005), Nr. 1, S. 50–55

BUM⁺04 BENYOUCEF, M ; ULRICH, S M. ; MICHLER, P ; WIERSIG, J ; JAHNKE, F ; FORCHEL, A: Enhanced correlated photon pair emission from a pillar microcavity. In: *New J. Phys.* 6 (2004), S. 91

BY99 BENSON, Oliver ; YAMAMOTO, Yoshihisa: Master-equation model of a single-quantum-dot microsphere laser. In: *Phys. Rev. A* 59 (1999), Jun, Nr. 6, S. 4756–4763

Car93 CARMICHAEL, H. J.: Quantum trajectory theory for cascaded open systems. In: *Phys. Rev. Lett.* 70 (1993), Nr. 15, S. 2273 – 2276

Car99 CARMICHAEL, H.J.: *Statistical Methods in Quantum Optics 1 - Master Equation and Fokker-Planck Equations*. Springer, Berlin Heidelberg New York, 1999

CG09 COISH, W. A. ; GAMBETTA, J. M.: Entangled photons on demand: Erasing which-path information with sidebands. In: *Phys. Rev. B* 80 (2009), Dec, Nr. 24, S. 241303

CKI94 CHOW, W. W. ; KOCH, S. W. ; III, M. S.: *Semiconductor Laser Physics*. Springer, 1994

CKR09 CARMELE, A. ; KNORR, A. ; RICHTER, M.: Photon statistics as a probe for exciton correlations in coupled nanostructures. In: *Phys. Rev. B* 79 (2009), S. 035316

CMD⁺10 CARMELE, A. ; MILDE, F. ; DACHNER, M.-R. ; HAROUNI, M. B. ; ROKNIKNIZADEH, R. ; RICHTER, M. ; KNORR, A.: Formation dynamics of an entangled photon pair – a temperature dependent analysis. In: *Phys. Rev. B.* 81 (2010), S. 195319

CRCK10 CARMELE, Alexander ; RICHTER, Marten ; CHOW, Weng W. ; KNORR, Andreas: Antibunching of Thermal Radiation by a Room-Temperature Phonon Bath: A Numerically Solvable Model for a Strongly Interacting Light-Matter-Reservoir System. In: *Phys. Rev. Lett.* 104 (2010), Apr, Nr. 15, S. 156801

CRD+10 CARMELE, A. ; RICHTER, M. ; DACHNER, M.-R. ; WOLTER, J. ; KNORR, A.: Theory of few photon dynamics in electrically pumped light emitting quantum dot devices. In: *Proc. of SPIE* 7597 (2010), S. 75971

CSS75 CHOW, W. W. ; SCULLY, M. O. ; STONER, J. O.: Quantum-Beat Phenomena described by Quantum Electrodynamics and Neoclassical Theory. In: *Phys. Rev. A* 11 (1975), Nr. 4, S. 1380–1388

CT99 COHEN-TANNOUDJI, C.: *Quantenmechanik*. Bd. 1. de Gruyter, 1999

CTDRG89 COHEN-TANNOUDJI, C. ; DUPONT-ROC, J. ; GRYNBERG, G.: *Photons and Atoms*. 1rd. Chichester : John Wiley & Sons, 1989

DAFK06 DANCKWERTS, J. ; AHN, K. J. ; FÖRSTNER, J. ; KNORR, A.: Theory of ultrafast nonlinear optics of Coulomb-coupled semiconductor quantum dots: Rabi oscillations and pump-probe spectra. In: *Phys. Rev. B* 73 (2006), Nr. 16, S. 165318

DMK05 DUC, Huynh T. ; MEIER, T. ; KOCH, S. W.: Microscopic Analysis of the Coherent Optical Generation and the Decay of Charge and Spin Currents in Semiconductor Heterostructures. In: *Phys. Rev. Lett.* 95 (2005), Aug, Nr. 8, S. 086606

DMR+10 DACHNER, Matthias-René ; MALIC, Ermin ; RICHTER, Marten ; CARMELE, Alexander ; KABUSS, Julia ; WILMS, Alexander ; KIM, Jeong-Eun ; HARTMANN, Gregor ; WOLTERS, Janik ; BANDELOW, Uwe ; KNORR, Andreas: Theory of carrier and photon dynamics in quantum dot light emitters. In: *Phys. Status Solidi B* 247 (2010), Nr. 4, S. 809–828

DWRK10 DANG, Thi Uyen-Khanh ; WEBER, Carsten ; RICHTER, Marten ; KNORR, Andreas: Influence of Coulomb correlations on the quantum well intersubband absorption at low temperatures. In: *Phys. Rev. B* 82 (2010), Jul, Nr. 4, S. 045305

ENSM80 EBERLY, J. H. ; NAROZHNY, N.B. ; SANCHEZ-MONDRAGON, J.J.: Periodic Spontaneous Collapse and Revival in a Simple Quantum Model. In: *Phys. Rev. Lett.* 44 (1980), Nr. 20, S. 1323–1326

EOSI04 EDAMATSU, Keiichi ; OOHATA, Goro ; SHIMIZU, Ryosuke ; ITOH, Tadashi: Generation of ultraviolet entangled photons in a semiconductor. In: *Nature* 431 (2004), Nr. 7005, S. 167–170

ERK+96 EFROS, Al. L. ; ROSEN, M. ; KUNO, M. ; NIRMAL, M. ; NORRIS, D. J. ; BAWENDI, M.: Band-edge exciton in quantum dots of semiconductors with a degenerate valence band: Dark and bright exciton states. In: *Phys. Rev. B* 54 (1996), Aug, Nr. 7, S. 4843–4856

FAD+02 FÖRSTNER, J. ; AHN, K. J. ; DANCKWERTS, J. ; SCHAARSCHMIDT, M. ; WALDMÜLLER, I. ; WEBER, C. ; KNORR, A.: Light Propagation- and Many-particle-induced Non-Lorentzian Lineshapes in Semiconductor Nanooptics. In: *Phys. Status Solidi B* 234 (2002), Nr. 1, S. 155–165

FAT+04 FASEL, Sylvain ; ALIBART, Olivier ; TANZILLI, Sebastien ; BALDI, Pascal ; BEVERATOS, Alexios ; GISIN, Nicolas ; ZBINDEN, Hugo: High-quality asynchronous heralded single-photon source at telecom wavelength. In: *New J. Phys.* 6 (2004), S. 163

FMR+09 FLAGG, E. B. ; MULLER, A. ; ROBERTSON, J. W. ; FOUNTA, S. ; DEPPE, D. G. ; XIAO, M. ; MA, W. ; SALAMO, G. J. ; SHIH, C. K.: Resonantly driven coherent oscillations in a solid-state quantum emitter. In: *Nat. Phys.* 5 (2009), Nr. 3, S. 203 – 207

FMWS97 FRICKE, J. ; MEDEN, V. ; WÖHLER, C. ; SCHÖNHAMMER, K.: Improved Transport Equations Including Correlations for Electron-Phonon systems: Comparison with Exact Solutions in One Dimension. In: *Ann. Phys.* 253 (1997), S. 177–197

Fri96 FRICKE, Jens: Transport Equations Including Many-Particle Correlations for an Arbitrary Quantum System: A General Formalism. In: *Ann. Phys.* 252 (1996), S. 479–498

FS90 FICK, E. ; SAUERMANN, G.: *The Quantum Statistics of Dynamic Processes.* Berlin : Springer, 1990

GBBL04 GATTI, A. ; BRAMBILLA, E. ; BACHE, M. ; LUGIATO, L. A.: Ghost Imaging with Thermal Light: Comparing Entanglement and Classical Correlation. In: *Phys. Rev. Lett.* 93 (2004), Aug, Nr. 9, S. 093602

Gla63 GLAUBER, R.J.: The Quantum Theory of Optical Coherence. In: *Phys.Rev.* 130 (1963), Nr. 6, S. 2529–2539

GMK06 GOLDE, Daniel ; MEIER, Torsten ; KOCH, Stephan W.: Microscopic analysis of extreme nonlinear optics in semiconductor nanostructures. In: *J. Opt. Soc. Am. B* 23 (2006), Nr. 12, S. 2559–2565

GSB95 GRUNDMANN, Marius ; STIER, Oliver ; BIMBERG, Dieter: InAs/GaAs pyramidal quantum dots: Strain distribution, optical phonons, and electronic structure. In: *Phys. Rev. B* 52 (1995), Oct, Nr. 16, S. 11969–11981

GWJ08 GIES, Christopher ; WIERSIG, Jan ; JAHNKE, Frank: Output Characteristics of Pulsed and Continuous-Wave-Excited Quantum-Dot Microcavity Lasers. In: *Phys. Rev. Lett.* 101 (2008), Nr. 6, S. 067401

GWLJ07 GIES, Christopher ; WIERSIG, Jan ; LORKE, Michael ; JAHNKE, Frank: Semiconductor model for quantum-dot-based microcavity lasers. In: *Phys. Rev. A* 75 (2007), Nr. 1, S. 013803

Hak94 HAKEN, Hermann: *Licht und Materie.* Bd. 2. 2. Auflage. Mannheim : B.I. Wissenschaftsverlag, 1994

HBG⁺01 HEITZ, Robert ; BORN, Harald ; GUFFARTH, Florian ; STIER, Oliver ; SCHLIWA, Andrei ; HOFFMANN, Axel ; BIMBERG, Dieter: Existence of a phonon bottleneck for excitons in quantum dots. In: *Phys. Rev. B* 64 (2001), Nov, Nr. 24, S. 241305

HHH⁺06 HENNESSY, K. ; HÖGERLE, C. ; HU, E. ; BADOLATO, A. ; IMAMOĞLU, A.: Tuning photonic nanocavities by atomic force microscope nano-oxidation. In: *Appl. Phys. Lett.* 89 (2006), Nr. 4, S. 041118

HK04 HAUG, H. ; KOCH, S. W.: *Quantum Theory of the Optical and Electronic Properties of Semiconductors.* Singapore : World Scientific, 2004

HMS⁺99 HEITZ, R. ; MUKHAMETZHANOV, I. ; STIER, O. ; MADHUKAR, A. ; BIMBERG, D.: Enhanced Polar Exciton-LO-Phonon Interaction in Quantum Dots. In: *Phys. Rev. Lett.* 83 (1999), Nr. 22, S. 4654–4657

Hoh10 HOHENESTER, Ulrich: Cavity quantum electrodynamics with semiconductor quantum dots: Role of phonon-assisted cavity feeding. In: *Phys. Rev. B* 81 (2010), Apr, Nr. 15, S. 155303

HPS07 HOHENESTER, U. ; PFANNER, G. ; SELIGER, M.: Phonon-Assisted Decoherence in the Production of Polarization-Entangled Photons in a Single Semiconductor Quantum Dot. In: *Phys. Rev. Lett.* 99 (2007), S. 047402

HSB03 HEITZ, Robert ; SCHLIWA, Andrei ; BIMBERG, Dieter: Exciton-LO-phonon coupling in self-organized InAs/GaAs quantum dots. In: *Phys. Status Solidi B* 237 (2003), Nr. 1, S. 308–319

HT56 HANBURY BROWN, R. ; TWISS, R.Q.: Correlations between Photons in two coherent beams of light. In: *Nature* 177 (1956), S. 27 – 32

HUM⁺07 HAFENBRAK, R ; ULRICH, S M. ; MICHLER, P ; WANG, L ; RASTELLI, A ; SCHMIDT, O G.: Triggered polarization-entangled photon pairs from a single quantum dot up to 30 K. In: *New Journal of Physics* 9 (2007), Nr. 9, S. 315

JC63 JAYNES, E.T. ; CUMMINGS, F.W.: Comparison of quantum and semiclassical radiation theories with application to the beam maser. In: *Proc.IEEE* 51 (1963), Nr. 1, S. 89–109

JKMW01 JAMES, Daniel F. V. ; KWIAT, Paul G. ; MUNRO, William J. ; WHITE, Andrew G.: Measurement of qubits. In: *Phys. Rev. A* 64 (2001), Oct, Nr. 5, S. 052312

Jos94 JOSZA, R.: Fidelity for mixed quantum states. In: *J. Mod. Opt.* 41 (1994), S. 2315–2323

KAK02 KRUMMHEUER, B. ; AXT, V. M. ; KUHN, T.: Theory of pure dephasing and the resulting absorption line shape in semiconductor quantum dots. In: *Phys. Rev. B* 65 (2002), May, Nr. 19, S. 195313

KGK+06 KHITROVA, G. ; GIBBS, H. M. ; KIRA, M. ; KOCH, S. W. ; SCHERER, A.: Vacuum Rabi splitting in semiconductors. In: *Nat. Phys.* 2 (2006), S. 81–90

KJHK99 KIRA, M. ; JAHNKE, F. ; HOYER, W. ; KOCH, S. W.: Quantum theory of spontaneous emission and coherent effects in semiconductor microstructures. In: *Prog. Quantum Electron.* 23 (1999), S. 189–279

KK06 KIRA, M. ; KOCH, S.W.: Many-body correlations and excitonic effects in semiconductor spectroscopy. In: *Prog. Quantum Electron.* 30 (2006), Nr. 5, S. 155–296

KK08 KIRA, M. ; KOCH, S.W.: Cluster-expansion representation in quantum optics. In: *Phys. Rev. A* 78 (2008), S. 022102

KK10 KALASHNIKOV, Dmitry A. ; KRIVITSKY, Leonid A.: Spectrally resolved quantum tomography of polarization-entangled states. In: *New Journal of Physics* 12 (2010), Nr. 9, S. 093040

KKKG06 KOCH, S. W. ; KIRA, M. ; KHITROVA, G. ; GIBBS, H. M.: Semiconductor excitons in new light. In: *Nat. Mater* 5 (2006), Nr. 523 - 531

KKM01 KOCH, S W. ; KIRA, M ; MEIER, T: Correlation effects in the excitonic optical properties of semiconductors. In: *Journal of Optics B: Quantum and Semiclassical Optics* 3 (2001), Nr. 5, S. R29

KMR+10a KIM, Jeong E. ; MALIC, E. ; RICHTER, M. ; WILMS, A. ; KNORR, A.: Maxwell-Bloch Equation Approach for Describing the Microscopic Dynamics of Quantum-Dot Surface-Emitting Structures. In: *Quantum Electronics, IEEE Journal of* 46 (2010), jul., Nr. 7, S. 1115 –1126

KMR+10b KISTNER, C. ; MORGENER, K. ; REITZENSTEIN, S. ; SCHNEIDER, C. ; HÖFLING, S. ; WORSCHECH, L. ; FORCHEL, A. ; YAO, P. ; HUGHES, S.: Strong coupling in a quantum dot micropillar system under electrical current injection. In: *Applied Physics Letters* 96 (2010), Nr. 22, S. 221102

KRM+10 KASPRZAK, J. ; REITZENSTEIN, S. ; MULJAROV, E.A. ; KISTNER, C. ; SCHNEIDER, C. ; STRAUSS, M. ; HOFLING, S. ; FORCHEL, A. ; LANGBEIN, W.: Up on the Jaynes-Cummings ladder of a quantum-dot/microcavity system. In: *Nat.Mater.* 9 (2010), S. 304

LMC08 LARSON, J. ; MOYA-CESSA, H.: Rabi Oscillations in a quantum dot-cavity system coupled to a nonzero temperature phonon bath. In: *Phys. Scr.* 77 (2008), S. 065704

LO05 LOUNIS, Brahim ; ORRIT, Michel: Single-photon sources. In: *Rep. Prog. Phys.* 68 (2005), S. 1129–1179

Lou83 LOUDON, R.: *The Quantum Theory of Light*. Oxford Univ. Press, 1983

LP07 LAMBROPOULOS, P. ; PETROSYAN, D.: *Fundamentals of Quantum Optics and Quantum Information*. Springer, 2007

LST+09 LOCHMANN, A. ; STOCK, E. ; TÖFFLINGER, J.A. ; UNRAU, W. ; TOROPOV, A. ; BAKAROV, A. ; HAISLER, V. ; BIMBERG, D.: Electrically pumped, micro-cavity based single photon source driven at 1 GHz. In: *Electron. Lett.* 45 (2009), Nr. 11, S. 566

LVT08 LAUSSY, F.P. ; VALLE, E. del ; TEJEDOR, C.: Strong Coupling of Quantum Dots in Microcavities. In: *Phys. Rev. Lett.* 101 (2008), Nr. 8, S. 083601

LVT09 LAUSSY, Fabrice P. ; VALLE, Elena del ; TEJEDOR, Carlos: Luminescence spectra of quantum dots in microcavities. I. Bosons. In: *Phys. Rev. B* 79 (2009), Jun, Nr. 23, S. 235325

Mah90 MAHAN, Gerald D.: *Many-Particle Physics*. New York : Plenum Press, 1990

MGJ01 MARQUEZ, J. ; GEELHAAR, L. ; JACOBI, K.: Atomically resolved structure of InAs quantum dots. In: *Appl. Phys. Lett.* 78 (2001), Nr. 16, S. 2309 – 2311

MIM+00 MICHLER, P. ; IMAMOGLU, A. ; MASON, M. D. ; CARSON, P. J. ; STROUSE, G. F. ; BURATTO, S. K.: quantum correlation among photons from a single quantum dot at room temperature. In: *Nature* 406 (2000), S. 968–970

MK00 MAY, Volkhard ; KÜHN, Olivier: *Charge and Energy Transfer Dynamics in Molecular Systems*. Berlin : Wiley-VCH Verlag, 2000

MKB+00 MICHLER, P. ; KIRAZ, A. ; BECHER, C. ; SCHOENFELD, W. V. ; PETROFF, P. M. ; ZHANG, Lidong ; HU, E. ; IMAMOGLU, A.: A Quantum Dot Single-Photon Turnstile Device. In: *Science* 290 (2000), Nr. 5500, S. 2282–2285

MKH08 MILDE, F. ; KNORR, A. ; HUGHES, S.: Role of electron-phonon scattering on the vacuum Rabi splitting of a single quantum dot and a photonic crystal nanocavity. In: *Phys. Rev. B* 78 (2008), S. 035330

MKZ+00 MICHLER, P. ; KIRAZ, A. ; ZHANG, Lidong ; BECHER, C. ; HU, E. ; IMAMOGLU, A.: Laser emission from quantum dots in microdisk structures. In: *Applied Physics Letters* 77 (2000), Nr. 2, S. 184–186

MR09 MACHNIKOWSKI, P. ; ROZBICKI, E.: Phonon-assisted excitation transfer in quantum dot molecules: from quantum kinetics to transfer rates. In: *Phys. Statuts Solidi B* 246 (2009), Nr. 2, S. 320–324

MSO⁺10 MALIK, Mehul ; SHIN, Heedeuk ; O'SULLIVAN, Malcolm ; ZEROM, Petros ; BOYD, Robert W.: Quantum Ghost Image Identification with Correlated Photon Pairs. In: *Phys. Rev. Lett.* 104 (2010), Apr, Nr. 16, S. 163602

MTK07 MEIER, T. ; THOMAS, P. ; KOCH, S. W.: *Coherent semiconductor optics - From Basic concepts to nanostructure applications*. Berlin : Springer, 2007

MW95 MANDEL, L. ; WOLF, E.: *Optical coherence and quantum optics*. Cambridge : Cambridge University Press, 1995

MZ07 MULJAROV, Egor A. ; ZIMMERMANN, Roland: Exciton Dephasing in Quantum Dots due to LO-Phonon Coupling: An Exactly Solvable Model. In: *Phys. Rev. Lett.* 98 (2007), Nr. 18, S. 187401

NBFZ06 NARVAEZ, Gustavo A. ; BESTER, Gabriel ; FRANCESCHETTI, Alberto ; ZUNGER, Alex: Excitonic exchange effects on the radiative decay time of monoexcitons and biexcitons in quantum dots. In: *Phys. Rev. B* 74 (2006), Nov, Nr. 20, S. 205422

NC00 NIELSEN, M. A. ; CHUANG, I. L.: *Quantum Computation and Quantum Information*. Cambridge : Cambridge University Press, 2000

NKI⁺10 NOMURA, M. ; KUMAGAI, N. ; IWAMOTO, S. ; OTA, Y. ; ARAKAWA, Y.: Laser oscillation in a strongly coupled single-quantum-dot-nanocavity system. In: *Nature Physics* 6 (2010), Nr. 4, S. 279

NRS⁺94 NORRIS, T. B. ; RHEE, J.-K. ; SUNG, C.-Y. ; ARAKAWA, Y. ; NISHIOKA, M. ; WEISBUCH, C.: Time-resolved vacuum Rabi oscillations in a semiconductor quantum microcavity. In: *Phys. Rev. B* 50 (1994), Nr. 19, S. 14663–14666

NSME81 NAROZHNY, N.B. ; SANCHEZ-MONDRAGON, J.J. ; EBERLY, J.H.: Coherence versus incoherence: Collapse and revival in a simple quantum model. In: *Phys. Rev. A* 23 (1981), Nr. 1, S. 236–246

PRS⁺07 PASENOW, Bernhard ; REICHELT, Matthias ; STROUCKEN, Tineke ; MEIER, Torsten ; KOCH, Stephan W.: Microscopic analysis of the optical and electronic properties of semiconductors. In: *phys. status solidi a* 204 (2007), Nr. 11, S. 3600–3617

RÖ02 RÖSSLER, Ulrich (Hrsg.): *Landolt-Börnstein - Group III Condensed Matter*. Bd. III/41b: *Group IV Elements, IV-IV and III-V Compounds*. Springer,, 2002

RAK+06 RICHTER, Marten ; AHN, Kwang J. ; KNORR, Andreas ; SCHLIWA, Andrei ; BIMBERG, Dieter ; MADJET, Mohamed El-Amine ; RENGER, Thomas: Theory of excitation transfer in coupled nanostructures - from quantum dots to light harvesting complexes. In: *Phys. Status Solidi B* 243 (2006), Nr. 10, S. 2302–2310

RBSK07 RICHTER, M. ; BUTSCHER, S. ; SCHAARSCHMIDT, M. ; KNORR, A.: Model of thermal terahertz light emission of a two-dimensional electron gas. In: *Phys. Rev. B* 75 (2007), S. 115331

RCB+09 RICHTER, M. ; CARMELE, A. ; BUTSCHER, S. ; BÜCKING, N. ; MILDE, F. ; KRATZER, P. ; SCHEFFLER, M. ; KNORR, A.: Two-dimensional electron gases: Theory of ultrafast dynamics of electron-phonon interactions in graphene, surfaces, and quantum wells. In: *J. Appl. Phys.* 105 (2009), Nr. 12, S. 122409

RCSK09a RICHTER, Marten ; CARMELE, Alexander ; SITEK, Anna ; KNORR, Andreas: Few-Photon Model of the Optical Emission of Semiconductor Quantum Dots. In: *Phys. Rev. Lett.* 103 (2009), Nr. 8, S. 087407

RCSK09b RICHTER, Marten ; CARMELE, Alexander ; SITEK, Anna ; KNORR, Andreas: Few-Photon Model of the Optical Emission of Semiconductor Quantum Dots. In: *Phys. Rev. Lett.* 103 (2009), Nr. 8, S. 087407

RGGJ10 RITTER, S. ; GARTNER, P. ; GIES, C. ; JAHNKE, F.: Emission properties and photon statisticsof a single quantum dot laser. In: *Opt. Express* 18 (2010), Nr. 10, S. 9909–9921

Rob03 ROBINETT, R. W.: Quantum wave packet revivals. In: *Phys. Rep.* 392 (2003), S. 1–119

RSL+04 REITHMAIER, J. P. ; SEK, G. ; LÖFFLER, A. ; HOFMANN, C. ; KUHN, S. ; REITZENSTEIN, S. ; KELDYSH, L. V. ; KULAKOVSKII, V. D. ; REINECKE, T. L. ; FORCHEL, A.: Strong coupling in a single quantum dot?semiconductor microcavity system. In: *Nature* 432 (2004), S. 197–200

RWK87 REMPE, G. ; WALTHER, H. ; KLEIN, N.: Observation of Quantum Collapse and Revival in a One-Atom Maser. In: *Phys. Rev. Lett.* 58 (1987), Nr. 4, S. 353 – 356

SBS06 SCARCELLI, Giuliano ; BERARDI, Vincenzo ; SHIH, Yanhua: Can Two-Photon Correlation of Chaotic Light Be Considered as Correlation of Intensity Fluctuations? In: *Phys. Rev. Lett.* 96 (2006), Feb, Nr. 6, S. 063602

Sch04 SCHOLES, Gregory D.: Selection rules for probing biexcitons and electron spin transitions in isotropic quantum dot ensembles. In: *J. Chem. Phys.* 121 (2004), Nr. 20, S. 10104–10110

SCK01 SCHNEIDER, H. C. ; CHOW, W. W. ; KOCH, S. W.: Many-body effects in the gain spectra of highly excited quantum-dot lasers. In: *Phys. Rev. B* 64 (2001), Nr. 11, S. 115315

SCK04 SCHNEIDER, H.C. ; CHOW, W.W. ; KOCH, S.W.: Excitation-induced Dephasing in Semiconductor Quantum Dots. In: *Phys. Rev. B* 70 (2004), S. 235308

SCR+10 SU, Y. ; CARMELE, A. ; RICHTER, M. ; LÜDGE, K. ; SCHÖLL, E. ; BIMBERG, D. ; KNORR, A.: Theory of single quantum dot lasers: Pauli-blocking enhanced anti-bunching. In: *Semicond. Sci. Technol.* (2010)

SDG+09 SUFFCZYSKI, J. ; DOUSSE, A. ; GAUTHRON, K. ; LEMAITRE, A. ; SAGNES, I. ; LANCO, L. ; BLOCH, J. ; VOISIN, P. ; SENELLART, P.: Origin of the Optical Emission within the Cavity Mode of Coupled Quantum Dot-Cavity Systems. In: *Phys. Rev. Lett.* 103 (2009), Jul, Nr. 2, S. 027401

SDT+05 SANVITTO, D. ; DARAEI, A. ; TAHRAOUI, A. ; HOPKINSON, M. ; FRY, P. W. ; WHITTAKER, D. M. ; SKOLNICK, M. S.: Observation of ultrahigh quality factor in a semiconductor microcavity. In: *Applied Physics Letters* 86 (2005), Nr. 19, S. 191109

SFP+02 SANTORI, Charles ; FATTAL, David ; PELTON, Matthew ; SOLOMON, Glenn S. ; YAMAMOTO, Yoshihisa: Polarization-correlated photon pairs from a single quantum dot. In: *Phys. Rev. B* 66 (2002), Jul, Nr. 4, S. 045308

SGB99a STIER, Oliver ; GRUNDMANN, Marius ; BIMBERG, Dieter: Electronic and optical properties of strained quantum dots modeled by 8-band k·p theory. In: *Phys. Rev. B* 59 (1999), S. 5688–5701

SGB99b STIER, Oliver ; GRUNDMANN, Marius ; BIMBERG, Dieter: Electronic and optical properties of strained quantum dots modeled by 8-band k·p theory. In: *Phys. Rev. B* 59 (1999), S. 5688–5701

SH06 STEEB, W.-H. ; HARDY, Yorick: *Problems and solutions in quantum computing and quantum information.* 2nd edition. Singapore : World Scientific, 2006

Shi07 SHIELDS, A. J.: Semiconductor Quantum Light Sources. In: *Nat.Photonics* 1 (2007), S. 215 – 223

SHY+85 SLUSHER, R. E. ; HOLLBERG, L. W. ; YURKE, B. ; MERTZ, J. C. ; VALLEY, J. F.: Observation of Squeezed States Generated by Four-Wave Mixing in an Optical Cavity. In: *Phys. Rev. Lett.* 55 (1985), Nov, Nr. 22, S. 2409–2412

SK93 SHORE, B. W. ; KNIGHT, P. L.: The Jaynes-Cummings-Model. In: *J. Mod. Opt.* 40 (1993), Nr. 7, S. 1195 – 1238

SKK08 SCHNEEBELI, L. ; KIRA, M. ; KOCH, S. W.: Characterization of Strong Light-Matter Coupling in Semiconductor Quantum-Dot Microcavities via Photon-Statistics Spectroscopy. In: *Phys. Rev. Lett.* 101 (2008), Nr. 9, S. 097401

SMO+02 SANGUINETTI, S. ; MANO, T. ; OSHIMA, M. ; TATENO, T. ; WAKAKI, M. ; KOGUCHI, N.: Temperature dependence of the photoluminescence of InGaAs/GaAs quantum dot structures without wetting layer. In: *Applied Physics Letters* 81 (2002), Nr. 16, S. 3067–3069

SRK+10 SU, Y. ; RICHTER, M. ; KNORR, A. ; BIMBERG, D. ; CARMELE, A.: Photon statistics of a single quantum dot in a microcavity. In: *Physica Status solidi (RRL)* 4 (2010), S. 289

SSF+10 SALTER, C. L. ; STEVENSON, R. M. ; FARRER, I. ; NICOLL, C. A. ; RITCHIE, D. A. ; SHIELDS, A. J.: An entangled-light-emitting diode. In: *Nature* 465 (2010), S. 594

SSR+05 SEGUIN, R. ; SCHLIWA, A. ; RODT, S. ; PÖTSCHKE, K. ; POHL, U. W. ; BIMBERG, D.: Size-Dependent Fine-Structure Splitting in Self-Organized InAs/GaAs Quantum Dots. In: *Phys. Rev. Lett.* 95 (2005), Nr. 25, S. 257402

STH+09 SOTIER, F. ; THOMAY, T. ; HANKE, T. ; KORGER, J. ; MAHAPATRA, S. ; FREY, A. ; BRUNNER, K. ; BRATSCHITSCH, R. ; LEITENSTORFER, A.: Femtosecond few-fermion dynamics and deterministic single-photon gain in a quantum dot. In: *Nat. Phys.* 5 (2009), Nr. 5, S. 352 – 356

Sti01 STIER, O. ; THOMSEN, C. (Hrsg.) ; BIMBERG, D. (Hrsg.) ; DÄHNE, M. (Hrsg.) ; RICHTER, W. (Hrsg.): *Berlin Studies in Solid State Physics. Bd. 7: Electronic and Optical Properties of Quantum Dots and Wires.* Berlin : Wissenschaft & Technik Verlag, 2001

STS+02 STEVENSON, R. M. ; THOMPSON, R. M. ; SHIELDS, A. J. ; FARRER, I. ; KARDYNAL, B. E. ; RITCHIE, D. A. ; PEPPER, M.: Quantum dots as a photon source for passive quantum key encoding. In: *Phys. Rev. B* 66 (2002), Aug, Nr. 8, S. 081302

SWL+09 SCHLIWA, Andrei ; WINKELNKEMPER, Momme ; LOCHMANN, Anatol ; STOCK, Erik ; BIMBERG, Dieter: In(Ga)As/GaAs quantum dots grown on a (111) surface as ideal sources of entangled photon pairs. In: *Phys. Rev. B* 80 (2009), Oct, Nr. 16, S. 161307

SYA+06 STEVENSON, R. M. ; YOUNG, R.J. ; ATKINSON, P. ; COOPER, K. ; RITCHIE, D.A. ; SHIELDS, A.J.: A semiconductor source of triggered entangled photon pairs. In: *Nature* 439 (2006), S. 179 – 182

SZ97 SCULLY, M. O. ; ZUBAIRY, M. S.: *Quantum Optics.* Cambridge : Cambridge University Press, 1997

Tak93 TAKAGAHARA, T.: Effects of dielectric confinement and electron-hole exchange interaction on excitonic states in semiconductor quantum dots. In: *Phys. Rev. B* 47 (1993), Feb, Nr. 8, S. 4569–4584

Tak00 TAKAGAHARA, T.: Theory of exciton doublet structures and polarization relaxation in single quantum dots. In: *Phys. Rev. B* 62 (2000), Dec, Nr. 24, S. 16840–16855

THM00 TROIANI, F. ; HOHENESTER, U. ; MOLINARI, E.: Exploiting exciton-exciton interactions in semiconductor quantum dots for quantum-information processing. In: *Phys. Rev. B* 62 (2000), S. R2263

TKH+09 TAWARA, T. ; KAMADA, H. ; HUGHES, S. ; OKAMOTO, H. ; NOTOMI, M. ; SOGAWA, T.: Cavity mode emission in weakly coupled quantum dot - cavity systems. In: *Opt. Express* 17 (2009), Nr. 8, S. 6643–6654

TPT06 TROIANI, F. ; PEREA, J.I. ; TEJEDOR, C.: Cavity-assisted generation of entangled photon pairs by a quantum-dot cascade decay. In: *Phys. Rev. B* 74 (2006), S. 235310

TS90 TEICH, M. C. ; SALEH, B. E. A.: Squeezed and antibunched light. In: *Phys. Today* 43 (1990), Nr. 6, S. 23

TS10 TAREL, G. ; SAVONA, V.: Linear spectrum of a quantum dot coupled to a nanocavity. In: *Phys. Rev. B* 81 (2010), Feb, Nr. 7, S. 075305

UFT07 URSIN, R. ; F. TIEFENBACHER, H. Weier T. Scheidl M. Lindenthal B. Blauensteiner T. Jennewein J. Perdigues P. Trojek B. Ömer M. Fürst M. Meyenburg J. Rarity Z. Sodnik C. Barbieri H. Weinfurter A. Z. T. Smitt-Manderbach: Entanglement-based quantum communication over 144 km. In: *Nat. Phys.* 3 (2007), Nr. 6, S. 481 – 486

UGA+07 ULRICH, S. M. ; GIES, C. ; ATES, S. ; WIERSIG, J. ; REITZENSTEIN, S. ; HOFMANN, C. ; LÖFFLER, A. ; FORCHEL, A. ; JAHNKE, F. ; MICHLER, P.: Photon Statistics of Semiconductor Microcavity Lasers. In: *Phys. Rev. Lett.* 98 (2007), S. 043906

USM+03 ULRICH, S. M. ; STRAUF, S. ; MICHLER, P. ; BACHER, G. ; FORCHEL, A.: Triggered polarization-correlated photon pairs from a single CdSe quantum dot. In: *Applied Physics Letters* 83 (2003), Nr. 9, S. 1848–1850

Vah03 VAHALA, K. J.: Optical microcavities. In: *Nature* 424 (2003), S. 839 – 846

VLT09 VALLE, Elena del ; LAUSSY, Fabrice P. ; TEJEDOR, Carlos: Luminescence spectra of quantum dots in microcavities. II. Fermions. In: *Phys. Rev. B* 79 (2009), Jun, Nr. 23, S. 235326

VSDS05 VALENCIA, Alejandra ; SCARCELLI, Giuliano ; D'ANGELO, Milena ; SHIH, Yanhua: Two-Photon Imaging with Thermal Light. In: *Phys. Rev. Lett.* 94 (2005), Feb, Nr. 6, S. 063601

VSSM+10 VASCONCELLOS S., Michaelis de ; S., Gordon ; M., Bichler ; T., Meier ; A., Zrenner: Coherent control of a single exciton qubit by optoelectronic manipulation. In: *Nature Photonics* 4 (2010), Nr. 8, S. 545

VUH+07 VOGEL, M. M. ; ULRICH, S. M. ; HAFENBRAK, R. ; MICHLER, P. ; WANG, L. ; RASTELLI, A. ; SCHMIDT, O. G.: Influence of lateral electric fields on multiexcitonic transitions and fine structure of single quantum dots. In: *Applied Physics Letters* 91 (2007), Nr. 5, S. 051904

VWW01 VOGEL, W. ; WELSCH, D. ; WALLENTOWITZ, S.: *Quantum Optics An Introduction*. 2nd edition. Berlin : WILEY-VCH, 2001

Wie09 WIERSIG, J. et. a.: Direct observation of correlations between individual photon emission events of a microcavity laser. In: *Nature* 460 (2009), S. 245

WKU$^+$05 WARD, M. B. ; KARIMOV, O. Z. ; UNITT, D. C. ; YUAN, Z. L. ; SEE, P. ; GEVAUX, D. G. ; SHIELDS, A. J. ; ATKINSON, P. ; RITCHIE, D. A.: On-demand single-photon source for 1.3 mu m telecom fiber. In: *Appl. Phys. Lett.* 86 (2005), Nr. 20, S. 201111

WLW09 WEBER, C. ; LINDWAL, G. ; WACKER, A.: Zero-phonon line broadening and satellite peaks in nanowire quantum dots: The role of piezoelectric coupling. In: *Phys. Statuts Solidi B* 246 (2009), Nr. 2, S. 337–341

WNIA92 WEISBUCH, C. ; NISHIOKA, M. ; ISHIKAWA, A. ; ARAKAWA, Y.: Observation of the coupled exciton-photon mode splitting in a semiconductor quantum microcavity. In: *Phys. Rev. Lett.* 69 (1992), Dec, Nr. 23, S. 3314–3317

Woo98 WOOTTERS, William K.: Entanglement of Formation of an Arbitrary State of Two Qubits. In: *Phys. Rev. Lett.* 80 (1998), Mar, Nr. 10, S. 2245–2248

WRI02 WILSON-RAE, I. ; IMAMOĞLU, A.: Quantum dot cavity-QED in the presence of strong electron-phonon interactions. In: *Phys. Rev. B* 65 (2002), May, Nr. 23, S. 235311

WSS$^+$09 WARMING, T. ; SIEBERT, E. ; SCHLIWA, A. ; STOCK, E. ; ZIMMERMANN, R. ; BIMBERG, D.: Hole-hole and electron-hole exchange interactions in single InAs/GaAs quantum dots. In: *Phys. Rev. B* 79 (2009), Mar, Nr. 12, S. 125316

WVT$^+$09 WINGER, Martin ; VOLZ, Thomas ; TAREL, Guillaume ; PORTOLAN, Stefano ; BADOLATO, Antonio ; HENNESSY, Kevin J. ; HU, Evelyn L. ; BEVERATOS, Alexios ; FINLEY, Jonathan ; SAVONA, Vincenzo ; IMAMOGLU, Atac: Explanation of Photon Correlations in the Far-Off-Resonance Optical Emission from a Quantum-Dot–Cavity System. In: *Phys. Rev. Lett.* 103 (2009), Nov, Nr. 20, S. 207403

YC05 YU, Peter Y. ; CARDONA, Manuel: *Fundamentals of Semiconductors*. Berlin : Springer, 2005

YKS$^+$02 YUAN, Zhiliang ; KARDYNAL, Beata E. ; STEVENSON, R. M. ; SHIELDS., Andrew J. ; LOB, Charlene J. ; COOPER, Ken ; BEATTIE, Neil S. ; RITCHIE, David A. ; PEPPER, Michael: Electrically Driven Single-Photon Source. In: *Science* 295 (2002), Nr. 5552, S. 102–105

YSH$^+$04 YOSHIE, T. ; SCHERER, A. ; HENDRICKSON, J. ; KHITROVA, G. ; GIBBS, H. M. ; RUPPER, G. ; ELL, C. ; SHCHEKIN, O. B. ; DEPPE, D. G.: Vacuum Rabi splitting with a single quantum dot in a photonic crystal nanocavity. In: *Nature* 432 (2004), S. 200–203

YTC00 YAMAMOTO, Yoshihisa ; TASSONE, Francesco ; CAO, Hui: *Semiconductor Cavity Quantum Electrodynamics*. Bd. 169. Berlin : Springer Tracts in Modern Physics, 2000

Die VDM Verlagsservicegesellschaft sucht für wissenschaftliche Verlage abgeschlossene und herausragende

Dissertationen, Habilitationen, Diplomarbeiten, Master Theses, Magisterarbeiten usw.

für die kostenlose Publikation als Fachbuch.

Sie verfügen über eine Arbeit, die hohen inhaltlichen und formalen Ansprüchen genügt, und haben Interesse an einer honorarvergüteten Publikation?

Dann senden Sie bitte erste Informationen über sich und Ihre Arbeit per Email an *info@vdm-vsg.de*.

Sie erhalten kurzfristig unser Feedback!

VDM Verlagsservicegesellschaft mbH
Dudweiler Landstr. 99   Telefon +49 681 3720 174
D - 66123 Saarbrücken   Fax   +49 681 3720 1749
www.vdm-vsg.de

Die VDM Verlagsservicegesellschaft mbH vertritt

Printed by Books on Demand GmbH, Norderstedt / Germany